T0074419

Cyber Arms
Security in Cyberspace

Stanislav Abaimov
Cyber Security and Electronic Engineering
University of Rome Tor Vergata Rome, Italy

Maurizio Martellini
Department of Science and High Technology
University of Uninsubria Como, Italy

CRC Press
Taylor & Francis Group
Boca Raton   London   New York

CRC Press is an imprint of the
Taylor & Francis Group, an **informa** business
A SCIENCE PUBLISHERS BOOK

CRC Press
Taylor & Francis Group
6000 Broken Sound Parkway NW, Suite 300
Boca Raton, FL 33487-2742

© 2020 by Taylor & Francis Group, LLC
CRC Press is an imprint of Taylor & Francis Group, an Informa business

No claim to original U.S. Government works

Version Date: 20200206

International Standard Book Number-13: 978-0-367-42495-4 (Hardback)

Visit the Taylor & Francis Web site at
http://www.taylorandfrancis.com

and the CRC Press Web site at
http://www.crcpress.com

Preface

In the era of covert cyber warfare, lethal autonomous weapon systems and an artificial intelligence arms race, cyber military operations can be carried across the world without the deployment of human operatives. AI-empowered offensive cyber tools are accessible globally, many of them are free of charge, and they progress faster than human ethical values.

Computerised communication was intended to boost human interaction, facilitate routine manual work, assist in data analysis and decision making globally, and to finally connect the entire world through one boundless network. The aggressive part of the human mind transformed cyberspace into a new battlefield, and made this evolutionary invention serve its goals. Cyber arms, having evolved from a piece of prank malicious code, turned into weapons of mass destruction, and are currently threatening human existence.

Obvious military advantages of stealthy cyber operations encourage countries to apply cyber weapons to gain strategic information, impact an enemy's command and control system, damage critical infrastructure, and disrupt early warning and other vital functions. And the accelerated development of sophisticated attack vectors, with frenetic acquisition of cyber arms, vulnerabilities and exploits, are similar to stockpiling conventional weapons of mass destruction.

Covert cyber offensive activities effectively serve the purposes of national and international espionage and sabotage. These rapidly evolving virtual capabilities are winning the contest due to the belated development of governing international laws and policies on safe and secure cyber environments and responsible state behaviour in cyberspace. A nation's cyber sovereignty, self-defence and countermeasures, attribution and investigation, and internationally binding norms still remain an apple of discord and the debates continue.

Overall accessibility of cyber arms, their dual-use nature, lack of effective global regulatory mechanisms and concerted enforcement measures, are creating the ground for a global catastrophe initiated through cyber space. With AI in action, the future cyber challenges appear to be even more threatening. The world is not ready to face the upcoming danger of global cyber wars.

Disclaimer

This is a unique project, combining the knowledge of a cybersecurity expert and a nuclear physicist. Opinions, conclusions and recommendations expressed or implied within are solely those of the authors and do not express any interest of any third party or agency.

This book is based on the recent real-life events and contains about six hundred links to publicly available documents. These references were made at the time when the links were included in the footnotes. However, it cannot be guaranteed that the content of the pages to which the links relate, or the link itself, are still the same at the time of reading. In this case, the titles of the articles and their authors' names will serve for guidance.

Contents

List of Figures and Tables

Figures

Tables

Abbreviations

ABM Treaty	Anti-Ballistic Missile Treaty
AI	Artificial Intelligence
ANN	Artificial Neural Network
APT	Advanced Persistent Threat
AWS	Autonomous Weapon Systems
CBRNe	Chemical, Biological, Radiological, Nuclear, explosives
CCI	Cyber Counterintelligence
CERT	Computer Emergency Response Team
CI	Critical Infrastructure
DES	Data Encryption Standard
DNS	Domain Name Server
DOD	Department of Defense
ECDLF	European Computer Driving Licence Foundation
EMP weapons	Electromagnetic Pulse Weapons
ICS	Industrial Control System
ICT	Information and Communications Technologies
IDS	Intrusion Detection System
IO	Information Operations
LAWS	Lethal Autonomous Weapon Systems
MAARS	Modular Advanced Armed Robotic System
ML	Machine Learning
MFA	Multi-factor Authentication
PLC	Programmable Logic Controller
ROM	Read Only Memory
SCADA	Supervisory Control and Data Acquisition
WMD	Weapon of Mass Destruction

Introduction

"As long as there are sovereign nations possessing great power, war is inevitable. That is not an attempt to say when it will come, but only that it is sure to come. That was true before the atomic bomb was made. What has changed is the destructiveness of war."

Albert Einstein

In the age of global communication networks, autonomously acting technologies and widely available AI-enhanced cyber tools, stealthy and covert cyber operations with unpredictable propagation have become an issue of significant global concern. Being well aware of the increasing military efficiency of cyber offensive activities and, undoubtedly, with a noble goal to protect their citizens, countries apply these techniques to gain strategic information, create an impact on the enemy's critical infrastructure, strategic logistics, command and control systems, early warning and other vital functions.

Following the overall interconnectivity and computerisation of the managerial environment, both the military and civil sectors have become fully dependent on automated systems. Their disruption is an ideal opportunity for state and non-state actors to implement a hostile operation with a highly disruptive impact at low cost. With their stealth nature and low risk of attribution, these attacks can be rightfully considered as an equivalent of military strikes.

The World Economic Forum's 2018 Global Risks Report identified cyber-security threats as one of the world's four most pressing challenges, along with environmental degradation, economic strains and geopolitical tensions. In its "Future Shocks" series, this report cautions against complacency and highlights the need to prepare for sudden and dramatic disruptions that can cause rapid and irreversible deterioration in the systems that the world relies on. It forecasted that cyber-attacks will constitute the third largest global threat in 2018.[1]

---

[1] WEF 2018 Global Risk Report, 13th edition, last retrieved on 16 July 2018 at http://www3.weforum.org/docs/WEF_GRR18_Report.pdf

The duality of cyber security tools, used both for defensive and offensive operations, creates an environment for cyber arms proliferation and misconduct; and their overall accessibility allows malicious actors to easily acquire and launch cyber arms. The UN Group of Governmental Experts on Developments in the Field of Information and Telecommunications in the Context of International Security[2] recognised the dual-use nature of Information and Communications Technologies (ICT) as either legitimate or malicious, as well as the security challenge posed by global interconnectedness and anonymity, and the hostile potential of state and non-state actors. Without the internationally approved definition from them, classification, licencing and regulation, the proliferation of cyber weapons is extremely hard to prevent.

The magnitude of the cyber threat is currently measured in hundreds of billions of dollars of the annual financial losses.[3] Cyber weapons can potentially be referred to as a new national deterrent, and weapons of mass destruction merged with cyber technologies can be seen as a new global danger. The non-proliferation of the use of cyber arms for offence is essential and should be under a legally binding global framework applied and enforced through multilateral institutions.

Due protection, early warning and effective response are especially crucial in chemical, biological, radioactive, nuclear and explosives (CBRNe) facilities, and extremely critical within nuclear weapons, whose damage, intentional or unintentional activation, not only entails country level process disruptions, but also endangers human existence globally. Though considered to be well protected and, in strategic areas, air-gapped, their automation and control systems are managed by computerised equipment and routinely updated software is vulnerable to cyber tools. Remote access to control systems, and even to a control facility, opens a channel to sophisticated cyber-attacks.

The potential impact of cyber threats increased exponentially when cyber technologies converged with the weapons of mass destruction and lethal autonomous weapon systems. As stated in the Report of the Center for the Study of Weapons of Mass Destruction on The Future of Weapons of Mass Destruction (2014), by 2030, "the proliferation of these weapons is likely to be harder to prevent and thus potentially more prevalent. Nuclear

---

[2] Report of the Group of Governmental Experts on Developments in the Field of Information and Telecommunications in the Context of International Security, A/68/98, United Nations, 68th session, 2013, last retrieved on 15 August 2018 at http://www.unidir.org/files/medias/pdfs/developments-in-the-field-of-information-and-telecommunications-in-the-context-of-international-security-2012-2013-a-68-98-eng-0-518.pdf

[3] The Economic Impact of Cybercrime – No Slowing Down, Executive Summary, McAfee, February 2018, last retrieved on 16 July 2018 at https://www.mcafee.com/enterprise/en-us/assets/executive-summaries/es-economic-impact-cybercrime.pdf

weapons are likely to play a more significant role in the international security environment, and current constraints on the proliferation and use of chemical and biological weapons could diminish. There will be greater scope for WMD terrorism, though it is not possible to predict the frequency or severity of any future employment of WMD. New forms of WMD – beyond chemical, biological, radiological, and nuclear weapons – are unlikely to emerge by 2030, but cyber weapons will probably be capable of inflicting such widespread disruption that the United States may become as reliant on the threat to impose unacceptable costs to deter large-scale cyber-attacks as it currently is to deter the use of WMD."[4]

The use of the cyber environment for conflicts has made it possible to blur the established differentiation between war and peace, by adding numerous subtle stages. Unpredictability, covert and stealthy characters, and the deterrent potential of cyber threats increase tension and push the states towards the cyber arms race escalation, which inevitably leads to cyber wars.

The goal of this book is to increase the global awareness on the emerging challenges of cyber offensive activities leading to a global catastrophe. Its major objectives are to provide an expert insight into current and future, malicious and destructive uses of the evolved cyber arms, and to support concerted global efforts in development of international cyber regulations.

This book is written for cyber security specialists, managers, government officials, politicians, experts developing international information security laws and global leaders. It analyses the cyber arms evolution from prank malicious codes into lethal weapons of mass destruction, raises awareness about the scale of cyber offensive conflicts, cyber warfare mutation, cyber arms race escalation and use of Artificial Intelligence for military purposes. It highlights the international efforts in cyber environment regulation, reviews the best practices of the leading cyber powers and their controversial approaches. It also proposes information security and cyber defence solutions and suggests definitions for selected conflicting cyber terms.

The disruptive potential of merging cyber tools with military weapons is examined from a technical point of view, as well as legal, ethical, and political angles. This project encourages the development and use of cyber technology up to a certain safe level of autonomy, within the sensible balance of benefits and dangers.

---

[4] John P. Caves, Jr. and W. Seth Carus, The Future of Weapons of Mass Destruction and Their Nature and Role in 2030, last retrieved on 9 June 2018 at http://ndupress.ndu.edu/Portals/97/Documents/Publications/Occasional%20Papers/10_Future%20of%20WMD.pdf

1   Cyber Offence Landscape

"O divine art of subtlety and secrecy!
Through you we learn to be invisible, through you inaudible;
and hence we can hold the enemy's fate in our hands."

The Art of War[1]

The invention of computerised communication revolutionised human life, enhanced human interaction, accelerated business management, and skyrocketed scientific research. Along with these benefits came the new unknown threats, stealthy cyber weapons and the fifth military domain.

Originally, malicious cyber codes were used for jokes and creative art projects. Over time, the application of malware revealed its offensive potential and attracted the attention of cyber-crime. Advanced cyber arms and weapons are now developed and produced by the flourishing cyber industry, and widely acquired and used by criminals, terrorists, national intelligence services, military forces, etc.

1.1 Cyber Arms Creation and Evolution

For the purpose of clarity in this book, we distinguish **cyber arms** – as software able to function as an independent agent, **cyber tools** – as cyber arms used for specific purposes, and **cyber-physical weapons** – to be physical weapons, i.e. robots, drones, with cyber command and control systems.

1.1.1 Cyber Arms Assortment

Cyberspace, an artificially created virtual reality, is inhabited by software able to implement certain functions. Specific pieces of code that can act as independent agents and run commands are called cyber arms. These have a dual-use nature and, depending on the aim of application, they can be used both as a cyber-security tool as well as a cyber-weapon.

The anthology below covers the evolution of selected malware, i.e. viruses, worms, trojans, bots, ransomware, implants, as well as

---

[1]   Sun Tzu, The Art of War, Canterbury Classics, 2016, book VI, Printers Row Publishing Group (PRPG), San Diego, USA, p. 13.

cryptominers, and highlights the increased scope of its destructive potential. Its history dates to the production of the first computer and cyberspace itself.

The end of the 1930s was marked by the birth of the first electric programmable computer, the Colossus, the 1940s – by the first digital computers, the 1950s – by computers capable of storing and running programs from memory and the first commercial scientific computers. Since that time, the computer race has only been escalating and the inquisitive human minds dedicated their efforts and intelligence to improve the new virtual reality and harness all its benefits.

The predecessors of modern cyberarms were harmless minimalistic programs that resembled games and sent cryptic messages to users. The "programmers" and fiction writers of 1970s–80s were enthusiastic in developing visionary ideas about the potential of this emerging virtual reality. The following examples illustrate four decades of cyber arms' evolution, from their "stone age" where they were "prank viruses" to the current state of AI-empowered malware frameworks.

1949

Over 20 years before the creation of the first malware, John von Neumann outlined the initial concepts of malicious software in "Theory of self-reproducing automata."[2]

1971

The "Creeper Virus", commonly recognised as the first computer virus, is released. Created as an experimental self-duplicating program to illustrate a mobile application, it located a computer in the network, copied itself inside the computer, simulated the printing process, destroyed the file, and displayed a message on the screen "I'm the creeper, catch me if you can!" It was also able to duplicate itself erasing previous versions. The piece of code named the "Reaper" was created to counter Creeper virus and, thus, became one of the first examples of an "anti-virus" program.

Two years later, Michael Crichton's movie "Westworld" screened the first concept of a computer virus, causing inexplicable problems and breakdowns spreading "like a disease" between the androids.[3]

1974

The "Rabbit" (or Wabbit) virus is developed. It overloaded the computer resources by multiplying itself until it blocked the system, reduced its

---

[2] Theory of self-reproducing automata, John von Neumann at http://cba.mit.edu/events/03.11.ASE/docs/VonNeumann.pdf, last retrieved on 14 June 2018

[3] IMDB synopsis of Westworld, last retrieved on 13 July 2018 https://www.imdb.com/title/tt0070909/plotsummary#synopsis

performance, and crashed the computer. Today, it would have been named a "fork bomb" as it caused a Denial-of-Service (DoS) attack.

John Brunner's novel "The Shockwave Rider"[4], published in 1975, coined the word "worm" to describe a self-replicating program that propagates itself through a computer network.

1978

The first "Trojan", a malicious computer programme "Animal", is released. It asked questions on the type of animal that the user was thinking about, while the related program "Pervade" was creating a copy of itself and of the "Animal" in every directory to which the current user had access. It was also able to replicate itself to other computers through copies in multi-user networks.

Both the computer systems and the operating systems evolved, and the viruses and network worms along with them. At the beginning of 1980s, the world had about ten various computer platforms. In parallel, several malicious code segments were born – Elk Cloner, Festering Hate, Blackout and Burp, etc.

1981

The "Elk Cloner" malware, the first "boot sector virus", was written by a 15-year old schoolboy, as a prank, for the Apple II computer, and placed on a floppy disk with a game. It was released at the fiftieth time the game was started, showing a blank screen with a poem about the virus. If a computer booted from an infected floppy disk, a copy of the virus was placed in the computer's memory. The virus copied itself into the boot sector of a floppy disk and then spread into all future disks.[5]

1983

The first "professionally designed" virus was developed as an experiment to be presented at a weekly seminar on computer security at the Lehigh University. The concept of "computer virus" was first introduced in this seminar by the virus author, Frederick Cohen, a computer scientist and inventor of virus defence techniques.[6] After eight hours of intense computation on a VAX 11/750 system running Unix, the virus was completed and ready for demonstration. The program could install itself in, or infect, the other elements of the system.

---

[4] J. Brunner, The Shockwave Rider, 1975, Harper and Row, San Fransico, USA.
[5] M. Rouse and Elk Cloner, September 2005, last retrieved on 14 July 2018 at https:// searchsecurity.techtarget.com/definition/Elk-Cloner
[6] A Computer Virus, 1984, Fred Cohen, last retrieved on 18 July 2018 at http://web.eecs. umich.edu/~aprakash/eecs588/handouts/cohen-viruses.html

1986

The first virus for the IBM-PC computers named Brain[7], able to infect the boot sector, is released. It was followed by Alameda (or "Yale"), Cascade, Jerusalem, Lehigh, and Miami (South African Friday the 13th), which were already able to infect the .COM and .EXE files and destroy them. "Cascade" was also the first encrypted virus with a hidden functionality and the ability to avoid detection.

In 1980s, two classes of computer viruses were distinguished:
- File infectors, that attached themselves to executable programs.
- Boot sector viruses, that took advantage of the special segment of executable code at the beginning of a floppy or hard disk which was executed by the PC on start-up.

While viruses were more common than worms initially, worms became a larger and more widespread threat (due to automated replication algorithms), following the expansion of computer networking.

1988

The Morris Worm is created and spreads rapidly throughout the world, becoming the first worm to spread extensively via Internet. After multiple infections, it slowed down the computer through overloading and collapsing it.

1989

The first polymorphic virus "1260" was created, changing its encryption with each successive infection. It resulted in the ineffectiveness of scanning for standard malware signatures.

The same year, the first stealth techniques emerged. These viruses could change or conceal their location in memory, actively avoiding detection and removal by anti-virus tools.

1990s

A new type of viruses was developed that used a polymorphic engine to mutate while keeping the initial algorithm unchanged, e.g., the "Whale" virus could rewrite its own instructions and changed itself in an unpredictable way. This made the current scanning antivirus techniques, which relied on fixed, predictable elements of each copy of the virus, outdated.

With the invention of the Internet in 1991, and its subsequent entry into every office and household, government and corporation, initiating

---

[7]   J. Layden, PC virus celebrates 20th birthday, 19 January 2006, last retrieved on 2 September 2018 at https://www.theregister.co.uk/2006/01/19/pc_virus_at_20/

an exchange of emails and electronic documents which soon became the most profitable target for malware. Floppy disks and programmes were no longer the leading cause of malware distribution. The expansion of the Internet led to its recognition as a new critical infrastructure and its protection became a national security issue.

1992

In March, the Michelangelo worm threatened (via text notifications) to cause extensive damage to computers around the world. Any floppy disk inserted into the infected system became infected as well. It was the first widespread cyber "time bomb" virus.

1993

Other cyber "time bomb", viruses "Leandro" or "Leandro & Kelly" and "Freddy Krueger", which were activated in October, spread quickly. In addition to floppy disks, these could be distributed through the Bulletin Board System and shareware. Those could enter a system via Microsoft Windows and MS-DOS vulnerabilities, and then impact completely unrelated operating systems, such as the very first GNU/Linux distributions.

A new category of malware – the multi-environmental macro virus Concept – appeared in cyberspace and became the dominant type of viruses until the end of 1990s.[8] Consisting of several Microsoft Word macros designed only to replicate, it contained user commands able to be stored, run as a group, spread and infect Microsoft Word documents which were now exchanged by users more often than programs.

The late 1990s–early 2000s can be described as years of a "global euphoria" caused by the expansive use of the Internet and a "gold rush" for malware evolution.

1999

Malware such as the Happy99, Melissa, ExploreZip and Kak worms was released. They spread much faster than their predecessors through Microsoft environments, invisibly attaching themselves to emails, displaying fireworks to hide the changes (Happy99), creating excessive network traffic (Melissa), destroying Microsoft Office documents (ExploreZed), etc.

2000

Within hours, the ILOVEYOU worm, also known as Love Letter, or Love Bug infected millions of Windows machines spreading through the

---

[8] What is Macro Virus? – Definition, Kaspersky Labs, last retrieved on 15 September 2018 at https://www.kaspersky.com/resource-center/definitions/macro-virus

attached "Love Letter". The quick spread of the tricky social engineering techniques raised the issue of the users' personal responsibility for spreading viruses. This malware based on similar vulnerabilities in the Microsoft office is still widely used.

During the same year, a DDoS attack "Project Rivolta" was launched by "Mafiaboy," a.k.a. 15-year-old Michael Calce, against Yahoo.com, which was the most popular search engine at the time.

2001

Worms Nimda, Sircam, Code Red, Klez et al. are released, benefiting from vulnerabilities created by previous strains. Their enhanced capacities allowed them to propagate themselves at a very high speed, and avoided detection using all possible means to spread themselves through emails, open network shared folders and drives, Internet Information Services (Microsoft IIS) vulnerabilities and web sites, e.g., Nimda.

With the new technological advancements, sophistication of information technologies, programming languages, and identification of vulnerabilities, malware is developed for the computers, mobile phones, photo frames, ATMs, and industrial control systems. It comes up from new and unexpected resources, competing in the speed of propagation, detectability, vitality, and even in behavioural intelligence.

2007

A targeted DDoS attack hits Estonia, disabling the Prime Minister's website as well as several government-owned organizations. Online services of Estonian banks, media outlets and government bodies were taken down by the increased levels of Internet traffic for three weeks. A massive number of spam emails were sent by botnets and a huge amount of automated online requests overwhelmed servers. As a result, ATMs and online banking services were temporarily unavailable; government employees were unable to communicate with each other by email; and newspapers and broadcasters could not deliver the news.[9]

Botnet
A botnet is a group of computers connected in a coordinated fashion for malicious purposes. Each computer in a botnet is called a bot. These bots form a network of compromised computers, which are controlled by a third party and used to transmit malware or spam, or to launch attacks.

Originally, botnets were created as a tool with valid purposes in the Internet relay chat (IRC) channels. Eventually, hackers exploited the vulnerabilities in IRC networks

---

[9] D. McGuinness, How a cyber attack transformed Estonia, BBC News, 27 April 2017, last retrieved on 14 July 2018 at https://www.bbc.com/news/39655415

> and developed bots to perform malicious activities such as password theft, keystroke logging, spam distribution, DDoS attacks, etc.
>
> *Technopedia*[10]

Storm Worm was identified as a fast spreading email spamming malware for the Microsoft operating systems, joining infected computers into the Storm botnet. By the end of June 2007, it had infected 1.7 million computers, and had compromised between 1 and 10 million computers during the two following months. It disguised itself as a news email containing a film about fake news stories and asked the user to download the attachment. Furthermore, the malware disguised itself as everything from video files to greeting cards, and attacks were continuously refreshed to coincide with holidays and current news event stories.[11]

2008

The computer worm "Conficker", also known as Downadup, infected about 10 million Microsoft server systems and is called one of the most serious infections ever seen.[12] Its victims were very high level targets, and included the German Bundeswehr[13], UK Ministry of Defence[14], Royal Navy warships and submarines[15], UK Parliament[16], French Navy[17], Sheffield Hospital, and Norwegian Police Service.

Without an operator guiding the malware, disabling security mechanisms and opening connections to receive instructions from a remote computer, the Conficker worm continued growing and created one of the largest botnets that had ever been recorded at the time.

---

[10] Botnet, Technopedia, last retrieved on 28 August 2018 at https://www.techopedia.com/definition/384/botnet

[11] P. Gutmann, "World's most powerful supercomputer goes online", 31 August 2007, last retrieved on 5 September 2018 at http://seclists.org/fulldisclosure/2007/Aug/520

[12] Barry Neid, CNN, $250K Microsoft bounty to catch worm creator, 13 February 2009, last retrieved on 14 July 2018 at http://edition.cnn.com/2009/TECH/ptech/02/13/virus.downadup/index.html

[13] Conficker-Wurm infiziert hunderte Bundeswehr-Rechner, 16 February 2009, last retrieved on 14 July 2018 at https://web.archive.org/web/20090321171256/http://www.pc-professionell.de/news/2009/02/16/conficker_wurm_infiziert_hunderte_bundeswehr_rechner

[14] L. Page, MoD networks still malware-plagued after two weeks, 20 January 2009, last retrieved on 14 July 2018 at https://www.theregister.co.uk/2009/01/20/mod_malware_still_going_strong,

[15] Ibid

[16] Ch. Arthur, House of Commons network hit by Conficker computer worm, 27 March 2009, last retrieved on 14 July 2018 at https://www.theguardian.com/technology/2009/mar/27/conficker-downadup-parliament-virus-april-1

[17] K. Wilsher, French fighter planes grounded by computer virus, 7 February 2009, last retrieved on 14 July 2018 at https://www.telegraph.co.uk/news/worldnews/europe/france/4547649/French-fighter-planes-grounded-by-computer-virus.html

2009

Malicious code, identified as W32.Dozer, is spread to destroy data on infected computers and to prevent these computers from being rebooted. A series of coordinated DoS attacks is launched against major government, news media, and financial websites in South Korea and the US. The estimated number of the hijacked computers varies widely; from 20,000 to 166,000 according to different sources[18,19]. Among the targeted websites were The White House, the US Ministry of Defense, the Ministry of Public Administration and Security, the National Intelligence Service, The Washington Post, the South Korean's President's home page, the South Korean National Assembly and US Forces Korea, etc. Some of the companies used in the attack were partially owned by several governments, complicating final attribution.

2010

Stuxnet, the next generation "computer virus that infiltrates highly secure computers not connected to the Internet" and the "first cyber-weapon of geopolitical significance" affected the Iran's Natanz enriching uranium nuclear facility through penetrating the system by running a Siemens programmable logic controller (PLC). It produced the most sophisticated attack of this time through manipulating the centrifuges to make them "self-destruct" over a long period of time, by causing the engine to rotate at the wrong speed to further damage the machines. It was specifically designed to attack the Siemens Simatic WinCC SCADA systems, installed in large facilities (like nuclear plants and utility companies) to manage operations.[20]

This incident is considered as "the first attack on critical industrial infrastructure that sits at the foundation of modern economies," notes The New York Times.[21] "The true end goal of Stuxnet is cyber sabotage. It's a cyberweapon basically," said a senior antivirus researcher at Kaspersky Labs.[22]

---

[18] Botnet worm in DoS attacks could wipe data out on infected PCs, 13 July 2009, last retrieved on 14 July 2018 at https://www.cnet.com/news/botnet - worm-in-dos-attacks-could-wipe-data-out-on-infected-pcs/

[19] UK, not North Korea, source of DDoS attacks, researcher says, Martyn Williams, IDG News Service, July 14, 2009, last retrieved on 14 July 2018 at https://www.computerworld. com/article/2526790/u-k---not-north-korea--source-of-ddos-attacks--researcher-says. html

[20] H. Stark, "Mossad's Miracle Weapon: Stuxnet Virus Opens New Era of Cyber War". Der Spiegel, 8 August 2011, last retrieved on 14 July 2018 at http://www.spiegel.de/ international/world/mossad-s-miracle-weapon-stuxnet-virus-opens-new-era-of-cyber-war-a-778912.html

[21] R. Richmond, malware Hits Computerized Industrial equipment, 24 September 2010, last retrieved on 15 October 2018 at https://bits.blogs.nytimes.com/2010/09/24/ malware-hits-computerized-industrial-equipment/

[22] Idem

2012

Flame (Flamer, Skywiper) is a modular computer malware that attacks computers running Microsoft Windows. Used for targeted cyber espionage in the Middle Eastern countries, it was announced by the MAHER Center of Iranian National Computer Emergency Response Team (CERT), Kaspersky Lab and CrySyS Lab of the Budapest University of Technology and Economics. CrySyS stated in their report that "sKyWIper is certainly the most sophisticated malware we encountered during their practice; arguably, it is the most complex malware ever found".[23]

Shamoon, a computer virus designed to target computer systems running Microsoft Windows in the energy sector, was announced the same year by Symantec, Kaspersky Lab, and Seculert. In August 2012, this virus wiped out the data and operating systems on thirty thousand computers connected to the network of the national oil firm of Saudi Arabia, Aramco.[24] Shamoon's code included a so-called kill switch, a timer set to attack at 11:08 a.m., the exact time that Aramco's computers were wiped of memory. The erasing mechanism was called Wiper.

In November 2016, the virus made a surprise comeback and was used in a fresh wave of attacks against targets in Saudi Arabia.[25] In December 2018, according to a press release, Saipem confirmed that they had experienced a cyberattack that involved a variant of the Shamoon malware.[26] The attack caused infrastructure and data availability issues, forcing the organization to carry out restoration activities. 300 systems on their network were crippled by the malware related to the 2012 Shamoon attacks.

The same name, Wiper, was given to an erasing component of Flame, a computer virus that attacked Iranian oil companies in May 2012. Iranian oil ministry officials claimed that the Wiper software code forced them to cut Internet connections to their oil ministry, oil rigs and the Kharg Island oil terminal, a conduit for 80 percent of Iran's oil exports.[27]

---

[23] Technical Report by Laboratory of Cryptography and System Security (CrySyS Lab) et al., v1.04, 30 May 2012, last retrieved on 17 October 2019 at https://www.webcitation.org/682bQ4f6J?url=http://www.crysys.hu/skywiper/skywiper.pdf

[24] N. Perlroth, In Cyberattack on Saudi Firm, U.S. Sees Iran Firing Back, 23 October 2012, last retrieved on 15 October 2018 at https://www.nytimes.com/2012/10/24/business/global/cyberattack-on-saudi-oil-firm-disquiets-us.html

[25] Shamoon: Back from the dead and destructive as ever, 30 November 2016, Symantec security response, at https://www.symantec.com/connect/blogs/shamoon-back-dead-and-destructive-ever

[26] Saipem: update on the cyber attack suffered, 12 December 2018, last retrieved on 13 January 2019 at http://www.saipem.com/sites/SAIPEM_en_IT/con-side-dx/Press%20releases/2018/Cyber%20attack%20update.page

[27] S. Dehghan, Iranian oil ministry hit by cyber-attack, 23 April 2012, last retrieved on 16 October 2018 at https://www.theguardian.com/world/2012/apr/23/iranian-oil-ministry-cyber-attack

2015

The power grid in Ukraine was attacked by the BlackEnergy3 malware, which caused the shutdown of 30 substations, thereby interrupting power supply to 230,000 people. A telephone denial-of-service attack prevented people from reporting outages to call centres.[28]

In January 2015, Wired magazine reported that a cyberattack on a steel mill in Germany had manipulated control systems in such a way that "a blast furnace could not be properly shut down, resulting in 'massive' – though unspecified – damage". The report, issued by the Germany's Federal Office for Information Security, indicates the attackers gained access to the steel mill through the plant's business network, and then successively worked their way into production networks to access systems controlling the plant's equipment. The attackers infiltrated the corporate network using a spear-phishing attack. Once the attackers got a foothold on one system, they further compromised a "multitude" of systems, including industrial components on the production network.[29]

This sophisticated malware can already intelligently perform undetectable covert functions, related to data theft, disruption of networks and computers, damaging control systems: it can adapt to environment and change behaviour. Some of the largest costs in this period relate to ransomware, a rapidly growing form of malware that locks targets out of their own data and demands a ransom in return for restoring access. The FBI reports that more than 4,000 ransomware attacks occur daily,[30] while other research sources state that 230,000 new malware samples were produced daily in 2015.[31]

2016

More than 60 strains of the Locky ransomware spread throughout Europe and infected several million computers. At the peak of the spread, over five thousand computers per hour were infected in Germany alone. This

---

[28] K. Zetter, "Inside the Cunning, Unprecedented Hack of Ukraine's Power Grid", Wired, 3 March 2016, last retrieved on 16 October 2018 at https://www.wired.com/2016/03/ inside-cunningunprecedented-hack-ukraines-power-grid/

[29] K. Zetter, "A Cyberattack Has Caused Confirmed Physical Damage for the Second Time Ever," Wired, 8 January 2015, last retrieved on 17 October 2018 at http://www.wired. com/2015/01/german-steel-mill-hack-destruction/. Original in German at Bundesamt für Sicherheit in der Informationstechnik, *Die Lage der IT-Sicherheit in Deutschland 2014* (Bonn, 2015), http://www.wired.com/wp-content/uploads/2015/01/Lagebericht2014. pdf.

[30] Ransomware Prevention and Response for CISOs, last retrieved on 17 October 2018 at https://www.fbi.gov/file-repository/ransomware-prevention-and-response-for-cisos. pdf/view

[31] 27% of all recorded malware appeared in 2015, 25 January 2018, retrieved on 18 October 2018 at https://www.pandasecurity.com/mediacenter/press-releases/ all-recorded-malware-appeared-in-2015/

malicious software is becoming more sophisticated and can additionally damage the system functionality. For example, the "Jigsaw" strain encrypts the data and then starts deleting groups of files to stimulate the victim to pay up quicker. "Chimera" threatens to post the victim's files online, including pictures and videos, if the ransom is not paid by the deadline. In 2016, the FBI estimated that Ransomware payments would reach 1 billion USD, compared with to the 24 million USD paid in 2015.[32]

A cyberattack on the SWIFT financial messaging network, allegedly through APT, led to the theft of 81 million USD from the central bank of Bangladesh.[33]

Tiny Banker Trojan (Tinba) infected more than 20 major banking institutions in the United States, including TD Bank, Chase, HSBC, Wells Fargo, PNC and Bank of America.[34] This trojan uses HTTP injection to force the user's computer to believe that it is on the bank's website. The fake page looks and functions just as the real one. The users enter their information to log on and Tinba indicates the bank webpage's "incorrect login information" return and redirects the user to the real website. Most modern banking trojans capture personal financial credentials and forward them both to the bank and to the criminals, creating the false sense of security in users.

Figure 1.1 from the EUROPOL 2017 report portrays the scheme of Banker Trojan operations[35]:

Figure 1.1: Simplified overview of the banking Trojan cybercrime value change

The Mirai botnet, composed primarily of embedded and IoT devices, overwhelmed several high-profile targets with unprecedented levels of network traffic.[36] It was called one of the most powerful and disruptive DDoS attacks seen to date, by infecting the Internet of Things devices,

[32] H. Weisbaum, Crime and Growing, NBC News, January 2017, last retrieved on 18 October 2018 at https://www.nbcnews.com/tech/security/ransomware-now-billion-dollar-year-crime-growing-n704646

[33] J. Finkle, SWIFT says hackers still targeting bank messaging system, 13 October 2017, last retrieved on 18 October 2018 at https://www.reuters.com/article/us-cyber-heist/swift-says-hackers-still-targeting-bank-messaging-system-idUSKBN1CI1DO

[34] Banking Trojans: From Stone Age to Space Era, 21 March 2017, EUROPOL, last retrieved on 27 October 2018 at https://www.europol.europa.eu/sites/default/files/documents/banking_trojans_from_stone_age_to_space_era.pdf

[35] Ibid

[36] M. Antonakakis et al, Understanding the Mirai Botnet, Usenix Security 2017, last retrieved on 27 October 2018 at https://elie.net/publication/understanding-the-mirai-botnet

such as digital cameras, wireless routers and DVR players. It resulted in the inaccessibility of several high-profile websites such as GitHub, Twitter, The Guardian, CNN, Reddit, Netflix, Airbnb and many others.[37] Researchers later determined that it infected nearly 65,000 devices in its first 20 hours, doubling in size every 76 minutes, and ultimately built a sustained strength of between 200,000 and 300,000 infections.[38]

2017

New trends appear in the cyber threat landscape, such as remote malicious document attachments, popularization of coin mining modules in major malware strains (e.g., Trick banking trojan), and advanced threats in social media channels.

As per the Proofpoint Q4 2017 Threat Report, "whether distributed with document attachments or via malicious links, in 2017 ransomware remained the top payload distributed in malicious messages, accounting for 57 percent of all malicious message volume. Rapid and wide fluctuations in cryptocurrency values also affected the ransomware market, with payment demands denominated in Bitcoin falling 73 percent quarter over quarter. Instead, attackers are increasingly setting ransom amounts in USD or local currency equivalents, although payments themselves are still generally made in Bitcoin or other major cryptocurrencies".[39]

The WannaCry ransomware attack spread globally in 2017. Exploits revealed in the NSA cyber arsenal leak of late 2016 were used to enable the propagation of the malware. WannaCry's reach came due to one of the leaked Shadow Brokers Windows vulnerabilities known as EternalBlue. As a result, a single weapon rendered more than 300,000 computer systems useless in at least 150 countries.[40] Beyond its financial cost, the WannaCry attack disrupted critical and strategic infrastructures across the world, including government ministries, railways, banks, telecommunications providers, energy companies, car manufacturers, and hospitals.

WannaCry illustrated a growing trend of using cyberattacks to target critical infrastructure and strategic industrial sectors. As per the Symantec

---

[37] DDoS attack that disrupted internet was largest of its kind in history, The Guardian, 26 October 2016, last retrieved on 18 October 2018 at https://www.theguardian.com/technology/2016/oct/26/ddos-attack-dyn-mirai-botnet

[38] G.M. Graff, How a dorm room Minecraft scam brought down the Internet, 13 December 2017, last retrieved on 27 October 2018 at https://www.wired.com/story/mirai-botnet-minecraft-scam-brought-down-the-internet/

[39] Proofpoint Q4 2017 Threat Report: Coin miners and ransomware are front and center, 17 January 2018, last retrieved 15 October 2018, https://www.proofpoint.com/us/threat-insight/post/proofpoint-q4-2017-threat-report-coin-miners-and-ransomware-are-front-and-center

[40] N. Whiting, Cyberspace Triggers a New Kind of Arms Race, 1 February 2018, last retrieved on 2 August 2018 at https://www.afcea.org/content/cyberspace-triggers-new-kind-arms-race

2018 Report, the vulnerabilities are becoming more difficult for attackers to identify and exploit; however there is an increase in attackers injecting malware implants into the supply chain to infiltrate organizations.[41]

Hijacking software updates provides attackers with an entry point for compromising well-protected targets. The Petya/NotPetya (Ranso.Petya) malware attack, June 2017, was the most notable case, which spread globally affecting Windows systems. Companies such as Merck, FedEx and Maersk each reported third-quarter losses of around 300 million USD as a result of NotPetya.[42]

In September, the Xafecopy Trojan attacked 47 countries, affecting Android operating systems. Kaspersky Lab identified it as malware from the Ubsod family, redirecting money transactions to the attackers' accounts through click based WAP billing systems. Attempts to use spear-phishing attacks were detected against companies operating nuclear power plants in the United States.[43]

As per the CISCO 2018 Annual Report, 50 percent of global web traffic was encrypted as of October 2017.[44] Encryption is meant to enhance security, but it also provides malicious actors with a powerful tool to conceal their hostile activity and inflict additional damage over time.

Leaked NSA Cyber Weapons[45]

The mysterious hacking group the Shadow Brokers in August 2016, claimed to have breached the spy tools of the elite NSA-linked operation known as the Equation Group. The Shadow Brokers offered a sample of alleged stolen NSA data and attempted to auction particularly significant alleged NSA tools, including a Windows exploit known as EternalBlue, which hackers have since used to infect targets in two high-profile ransomware attacks.

WikiLeaks claims that Vault 7 reveals "the majority of [the CIA] hacking arsenal including malware, viruses, trojans, weaponised 'zero day' exploits, malware remote control systems and associated documentation." Experts agree that the leaks caused major problems for the CIA, both in terms of how the agency is viewed by the public and in its operational capabilities. And as with the Shadow Brokers releases, Vault 7 has led to a heated debate about the problems and risks inherent in the government's development of cyber arms.

---

[41] Symantec, Internet Threat Report 2018, p. 5
[42] C. Forrest, "NotPetya Ransomware Outbreak Cost Merck More Than $300M per Quarter", TechRepublic, 30 October 2017, last retrieved on 15 October 2018 at https://www.techrepublic.com/article/notpetya-ransomwareoutbreak-cost-merck- more-than-300m-per-quarter/
[43] S. Gallagher, "FBI-DHS "Amber" Alert Warns Energy Industry of Attacks on Nuke Plant Operators". Ars Technica, 7 July 2017, last retrieved on 26 October 2018 at https://arstechnica.com/information-technology/2017/07/dhs- fbi-warn-ofattempts-to-hack-nuclear-plants/
[44] 2018 Annual Cybersecurity Report, CISCO, last retrieved on 15 December 2018 at https://www.cisco.com/c/en/us/products/security/security-reports.html
[45] Vault7, WikiLeaks, https://wikileaks.org/ciav7p1/

The year 2017 witnessed a rise of AI-powered cyberattacks, where machine learning is used to study patterns of traffic flow and mimic legitimate network traffic, hide payload from antimalware systems, etc.[46]

The 2017 Presidential campaign in France

Two days prior to the presidential runoff in France in May 2017, unknown hackers released 9GB of leaked emails from the party of the front-runner candidate Emmanuel Macron. The leak seemed to aim to give Macron minimal time and ability to respond, since French presidential candidates are barred from speaking publicly starting two days before an election. But the Macron campaign did release statements confirming that the *En Marche!* party had been breached, while cautioning that not everything in the data dump was legitimate. After the email leak heading into the election, the Macron campaign said in a statement, "Intervening in the last hour of an official campaign, this operation clearly seeks to destabilize democracy, as already seen in the United States' last presidential campaign. We cannot tolerate that the vital interests of democracy are thus endangered."[47]

2018

One of the SAMSAM ransomware strains compromised the networks of multiple US victims, including the previous years attacks on healthcare facilities that were running out-dated versions of application. In March 2018, the Atlanta's city government announced that it was experiencing outages on various customer facing applications, including some that customers might use to pay bills or access court-related information. A city employee published a screenshot of a ransomware message demanding a payment of 51,000 USD to unlock the entire system.

The same year, security researchers discovered a new feature of the Rakhni trojan that changes behaviour and installs either a ransomware or cryptocurrency miners on an infected system based on the available computational resources. It spreads via phishing emails, and infections were detected in Russia, Kazakhstan, Ukraine, Germany, and India.[48] This malware is among the few, that have the capability to adapt and to counter the modern anomaly-based intrusion detection systems, since the malware changes its behaviour it they spreads.

---

[46] D. Dutt, 2018: the year of the AI-powered cyberattack, 10 January 2018, last retrieved on 2 August 2018 at https://www.csoonline.com/article/3246196/cyberwarfare/2018-the-year-of-the-ai-powered-cyberattack.html

[47] A. Greenberg, Hackers hit macron with huge email leak ahead of French election, 5 May 2017, last retrieved on 15 July 2018 at https://www.wired.com/2017/05/macron-email-hack-french-election/

[48] Adapting to the Times: Malware decides Infection, probability with Ransomware or Coinminer, 9 July 2018, last retrieved on 14 July 2018 at https://www.trendmicro.com/vinfo/se/security/news/cybercrime-and-digital-threats/adapting-to-the-times-malware-decides-infection-profitability-with-ransomware-or-coinminer

Rakhni Trojan

If the system has a cryptocurrency wallet installed, the malware infects the system with ransomware (RANSOM_RAKHNI.A). However, if it does not find a wallet and detects that the system has more than two processors, it downloads a miner (Coinminer_ MALBTC.D-WIN32) and remotely exploits the systems' resources. It uses Minergate and installs fake root certificates to mine Monero, Monero Original or Dashcoin cryptocurrencies. The user will observe a noticeable slowdown as Rakhni terminates processes of known applications. The researchers also observed a worm component that allows it to copy itself to all computers found in the local network, as well as the ability to disable Windows Defender if the systems scan shows no antivirus installed and simultaneously infect the entire system with spyware.

Kaspersky Lab [49]

In May 2018, a new malware, known as VPNFilter, targeted a range of routers and network-attached storage devices and could damage the operating system of the infected devices. Unlike most other IoT threats, it was capable of maintaining a persistent presence on an infected device, even after a reboot and its diverse capabilities include spying on traffic being routed through the device. In addition, this malware has a module which specifically intercepts Modbus SCADA communications. [50]

In July 2018, a Ukrainian intelligence agency claimed it had stopped a cyberattack against a chlorine plant that was launched using the VPNFilter malware. The attack allegedly aimed to disrupt the stable operation of the plant, which provided sodium hypochlorite (aka liquid chlorine) for water treatment. [51]

In 2018–2019, malware has become more advanced and harder to combat. The world faces threats from network-based ransomware worms to devastating wiper malware, enhanced by AI. At the same time, adversaries are getting more adept at creating malware that can evade traditional sandboxing. [52]

The shift towards machine learning and AI is the next evolutionary step which has already brought many beneficial applications, ranging from machine translation to image analysis. However, AI systems have their own weaknesses and vulnerabilities. Thus, image recognition algorithms are susceptible to "pixel poisoning" an attack that leads to

---

[49] Rakhni Trojan: To encrypt and to mine, 6 July 2018, last retrieved on 15 July 2018 at https://www.kaspersky.com/blog/rakhni-miner-cryptor/22988/

[50] VPNFilter: New Router malware with Destructive Capabilities, 23 May 2018, last retrieved on 15 October 2018 at https://www.symantec.com/blogs/threat-intelligence/vpnfilter-iot-malware

[51] J. Leyden, Ukraine claims it blocked VPNFilter attack at chemical plant, 13 July 2018, last retrieved on 14 October 2018 at https://www.theregister.co.uk/2018/07/13/ukraine_vpnfilter_attack/

[52] 2018 Annual Cybersecurity Report, last retrieved 15 July 2018 at https://www.cisco.com/c/en/us/products/security/security-reports.html

classification problems.[53] Algorithms trained using open-source data could be particularly vulnerable to this challenge as adversaries attempt to corrupt the data that other countries might use to train algorithms for military purposes. Those sophisticated cyberattacks could also lead to the exploitation of defence algorithms trained on more secure networks.

A new generation of AI malware, like DeepLocker, demonstrates enhanced evasive capabilities. It hides the malicious payload in carrier applications, such as video conference software, to avoid detection by most antivirus and malware scanners. Automated techniques make it easier to carry out large-scale attacks that require extensive human labour, e.g. "spear phishing," which involves gathering and exploiting personal data of victims. Researchers at ZeroFox demonstrated that a fully automated spear phishing system could create tailored messages on Twitter based on a user's demonstrated preferences, achieving a high rate of clicks to a link that could be malicious.[54]

The 2018 report "The Malicious Use of Artificial Intelligence: Forecasting, Prevention, and Mitigation"[55], released by a team of academics and researchers from the Oxford University, Cambridge University, Stanford University, the Electronic Frontier Foundation, artificial intelligence research group OpenAI, and other institutes, expresses their serious concerns. The authors conclude that artificial intelligence and machine learning are altering the landscape of security risks for citizens, organizations, and states. The malicious use of AI could threaten digital security (e.g., through the criminals training machines to hack or socially engineered victims at human or superhuman levels of performance), physical security (e.g., non-state actors weaponizing consumer drones), and political security (e.g., through privacy-eliminating surveillance, profiling, and repression, or through automated and targeted disinformation campaigns).

The report concludes that as AI capabilities become more advanced and widespread, the growing use of AI systems will lead to the following changes in the landscape of threats:

- *Expansion of existing threats.* The costs of attacks may be lowered by the scalable use of AI systems to complete tasks that would ordinarily

---

[53] X. Chen, C. Liu, B. Li, K. Lu, D. Song, Targeted Backdoor Attacks on Deep Learning Systems Using Data Poisoning, 2017, last retrieved 15 July 2018 at https://arxiv.org/pdf/1712.05526.pdf

[54] J. Seymour and P. Tully, Weaponizing data science for social engineering: automated E2E spear phishing on Twitter, 2016, last retrieved 15 July 2018 at https://www.blackhat.com/docs/us-16/materials/us-16-Seymour-Tully-Weaponizing-Data-Science-For-Social-Engineering-Automated-E2E-Spear-Phishing-On-Twitter-wp.pdf

[55] Report on The Malicious Use of Artificial Intelligence: Forecasting, Prevention and Mitigation, February 2018, last retrieved 15 July 2018 at http://img1.wsimg.com/blobby/go/3d82daa4-97fe-4096-9c6b-376b92c619de/downloads/1c6q2kc4v_50335.pdf

require human labour, intelligence and expertise. A natural effect would be to expand the set of actors who can carry out specific attacks, the rate at which they can carry out these attacks, and the set of potential targets.

- *Introduction of new threats.* New attacks may arise using AI systems to complete tasks that would be otherwise impractical for humans. In addition, malicious actors may exploit the vulnerabilities of AI systems deployed by defenders.
- *Change to the typical character of threats.* There is reason to expect attacks enabled by the growing use of AI to be especially effective, finely targeted, difficult to attribute, and likely to exploit vulnerabilities in AI systems.[56]

2019

On 4 April 2019 the American, National Security Agency Research Directorate released the source code of Ghidra, software reverse engineering (SRE) suite of tools in support of the Cybersecurity mission. This source code repository includes instructions to build on all supported platforms (macOS, Linux, and Windows). Developers will be able to collaborate by creating patches, and extending the tool to fit their cybersecurity needs. Ghidra is one of many open source software (OSS) projects developed within the National Security Agency.[57] Complete source code for Ghidra along with build instructions is available in the NSA public repository.[58]

In support of NSA's Cybersecurity mission, Ghidra was built to solve scaling and teaming problems on complex SRE efforts, and to provide a customizable and extensible SRE research platform. NSA has applied Ghidra SRE capabilities to a variety of problems that involve analysing malicious code and aid reverse engineering analysts to better understand potential vulnerabilities of software.[59]

1.1.2 Cyber Offence Tools and Techniques

The modern cyber offensive tools are numerous. In 2018, certain offensive operating systems comprised up to 1700 cyber offensive and defensive tools, scripts and databases, pre-installed and pre-configured to be used right after the deployment.[60]

---

[56] Ibid
[57] https://code.nsa.gov/
[58] Ghidra, NSA's Research Directorate, https://ghidra-sre.org/
[59] Ghidra Software Reverse Engineering Framework, NationalSecurityAgency/ghidra, GitHub, https://github.com/NationalSecurityAgency/ghidra
[60] Top 10 Linux Distributions for Ethical hacking and Penetration Testing, last retrieved on 28 October 2018 at https://resources.infosecinstitute.com/top-10-linux-distro-ethical-hacking-penetration-testing/

For the purpose of simplification and visibility, the classification of publicly and commercially available cyber tools may follow the same logic as an average penetration test: reconnaissance (intelligence gathering), scanning and vulnerability assessment, initial breach, privilege escalation, assault, persistence, exfiltration, covering tracks (Table 1.1). This classification is not exhaustive, but in general it covers all major areas of cyber tools application as per the cyberattack development scenario.

In legal contractual vulnerability assessments and penetration testing, the same tools and techniques are used, however with more caution and multiple constraining rules and limits.

Given below is a short description of the application of cybertools and some specific characteristics.

Reconnaissance

At the initial stage the goal is to collect as much information about the target as possible without directly interacting with the target systems and network.

Reconnaissance tools are used to gather information about the target, its networks and systems. This may include gathering information from public websites, Domain Name Server (DNS) records, collecting metadata from accessible documents, retrieving specific information through search engines, or any other such similar activities.

Detailed information about companies can be acquired from websites, technical job announcements, etc. For example, a company may announce a position opening for the network administrator with knowledge of Windows Server 2013, Unix-server administration, MySQL database management, Cisco equipment support, etc.

Network reconnaissance (slang "Network sniffing") is performed through passive traffic collection. Physical reconnaissance includes social engineering and physical observations of the target operation mode. Wireless network traffic can be passively captured in close physical proximity without connecting to the access point of the target.

Long-term reconnaissance (surveillance) can be conducted by governments at the national level to ensure internal control against crime, terrorism, and external threats, e.g., telephone surveillance, voice over IP traffic analysis, traffic metadata collection.

Snowden's revelation

In 2013, news reports in the international media revealed operational details about the United States National Security Agency (NSA) and its international partners' global surveillance of foreign nationals and US citizens. The reports mostly emanate from a cache of top secret documents leaked by ex-NSA contractor Edward Snowden, which he obtained while working for Booz Allen Hamilton, one of the largest contractors for defence and intelligence in the United States. In addition to a trove of US federal documents, Snowden's cache reportedly contains thousands of Australian, British and Canadian intelligence files that he had accessed via the exclusive "Five Eyes" network. These media reports have shed light on the implications of several secret treaties signed by members of the UK/USA community in their efforts to implement global surveillance.

The Washington Post[61]

The best defence against reconnaissance is to limit disclosed information, i.e. limit the outgoing data from the organization itself. For example, one of the oldest best practices is to disable *zone transfer* to "unknown" computers, so as not to disclose information about the local subdomains.

Scanning and Vulnerability Assessment

After the reconnaissance phase is complete, and the information about the target is passively collected, for additional system-specific information, the attackers will scan the network access points. Scanning tools collect information about the perimeter computer systems for potential entry points. They include port scanners, vulnerability analysis tools, operating system fingerprinting tools, banner grabbing tools, etc., which enumerate the infrastructure devices, networks, and systems in place, assess vulnerabilities and open ports and services operating on those ports, and fingerprint operating systems.

Scanning is done by performing a thorough check of the network range to identify active IP addresses, types of scans filtered by firewalls, active and potentially vulnerable processes and programs. Several tools can be used at this stage. For example, Hping3 is the manual packet crafting tool for protocol exploitation and network scanning. Nmap is known to be the most popular automated port scanner. Masscan is used as an alternative to Nmap due to its increased performance over large address ranges.

The best defence against scans is to filter information exposed to scanners, by limiting the software type and version details during the initial connection request. Many scanning tools depend on services running on common ports and open access to information to generate their banners. For instance, if the scanner detects a port 21 in an open state, it will typically assume that the service behind it is using FTP protocol. Changing these basic configurations may confuse the attackers. Furthermore, certain scripts allow for the creation of the illusion that all 65535 ports are open, overwhelming the attackers scanning software with

---

61 Edward Snowden, after months of NSA revelations says his mission's accomplished, 23 December 2013, last retrieved on 15 July 2018 at https://www.washingtonpost.com/world/national-security/edward-snowden-after-months-of-nsa-revelations-says-his-missions-accomplished/2013/12/23/49fc36de-6c1c-11e3-a523-fe73f0ff6b8d_story.html?noredirect=on&utm_term=.aafd374349b0

Table 1.1: Classification of cyber tools

No	Offensive tools	Function	Methods	Examples
1.	Reconnaissance tools	Information gathering for further planning and targeting	Google hacking Social Media Official Websites DNS services Metadata	Domain names Metadata Search engine that specializes in providing information on individuals Search engine that provides information on web forms Whois and Domain Name Servers Nslookup Metagoofil Exiftool Maltego Recon-NG Sublistr
2.	Scanning and vulnerability assessment tools	Network mapping and processes enumeration Vulnerability discovery	Ping sweep Banner grabbing Traffic interception	Nmap – a port scanner to detect processes, vulnerabilities, fingerprint operating systems, etc. Nessus, OpenVAS - a vulnerability scanning tool Burp Suite, ZAP – web application hacking suites Cain & Abel – advanced network scanning tool and network protocol manipulation Wireshark – traffic sniffing and analysis tool
3.	Breach and exploitation tools	Gain access to at least one active computer system in the network and escalate privileges	Password guessing Exploitation Phishing Rogue devices	Password attack tools (Hydra, John the Ripper) Metasploit Framework CANVAS Professional Manual exploitation Post exploitation tools and exploits

(Contd.)

4.	Assault tools	Damage or disrupt, change the configurations or environment variables of the compromised system. Cause DoS or false readings	Through software or hardware	Post exploitation tools and exploits, as well as built-in or common operating system utilities (date, time changes) Manual configuration file manipulation System file removal
5.	Exfiltration tools	Hide or extract data over common protocols to remove or copy it from a compromised system	Covert channels Cloud services Physical device extraction	OpenVPN OpenSSH FTP Tor TrueCrypt
6.	Persistence tools	Sustain operation by maintaining persistent presence in the network	Malware Backdoors Stealth user accounts	Netcat SSH server Tor hidden service
7.	Obfuscation tools	Conceal the presence when operating in the system or network, and origins of the attack	Using intermediary computers and devices to reroute ("bounce")	Proxy Server VPN Server and Client Tor tunnelling ("obfuscation proxy", obfs4)
8.	Covering tracks tools	Remove traces of presence from the network and computers	Manual or automated command execution Permanent Denial of Service	Log deletion (manual or automated)

false positives, thus the attacker does not know which services are actually active. Instead, in this case the attacker can only send requests to all the open ports, causing a potential DoS.

From the attacker's perspective, to avoid detection, scanning should be performed from different origins, e.g. different networks, geographically disparate locations, with changes of the language and local time zone of the operating system, to avoid potential detection.

Breach and Exploitation Tools

After the scanning phase is complete, and the potential vulnerabilities identified, the attackers gain access and escalate privileges. The breach and exploitation tools are used to manually, or automatically, execute malicious code against the vulnerable processes and services in the target system, deliver and deploy payloads (*msfvenom, veil*, etc.) in the network services. Payload is a piece of code and a part of exploit that is executed through the vulnerability. It can be a command, several commands, or a shell code (interactive remote command line interface with the target system).

Exploitation toolkits, such as Metasploit Framework, come preinstalled with most of security testing operating systems (e.g., Kali Linux, Parrot Security, BackBox, etc.). Exploits against the most well-known and popular vulnerabilities have been unified in the Metasploit Framework to be easily launched via a simple interface, allowing even unskilled users to carry out basic cyberattacks against vulnerable systems.

Exploitation provides the same level of the user privileges as the vulnerable application. Exploiting web applications provides attackers with minimal privileges and limited access to the system, while exploiting file-sharing protocols (e.g., SMB) can result in higher privileges.

Defences against the breach and exploitation tools can be ensured through cyber hygiene and well-written and validated program codes.

Assault Tools

Assault techniques and tools change configurations or environment variables of the system, and target both software and hardware.

A number of simple and automated tools exist to allow the attackers to achieve the assault effect (e.g., "fork bomb" command, scripts to remove system files, etc.). For example, altering operating system settings and changing permissions of critical directories can cause confusion in sophisticated software for system administrators and at the program level. Changing the geographical location settings and time zone (e.g., using *tzutils*) can modify the way the software calculates time, stores event history, sends emails, creates entries in calendars, etc. Changing permissions of the directory (e.g., *umask, chmod,* etc.) to "write-only" will not allow the software to "read" data from the configuration files. Changing the environmental variable in the operating system can cause software to address malicious scripts and trigger malware code by running basic system commands.

Assault techniques can be used involuntarily even by unprivileged users by initiating long running processes, e.g., as a result of a program being improperly terminated (process seems closed, but still uses resources), through opening multiple applications and tabs which engage large amounts of system resources such as memory, CPU, and storage drive. This prevents other systems or network processes from accessing them, creating a local DoS attack.

Physical destruction of equipment can be performed, for example, through attacking a Read Only Memory (ROM) module that consists of electronically reprogrammable memory. Using universal ROM flashing tools (e.g., flashrom), it is possible to rewrite the contents of such modules to alter the functionality of the hardware. Those tools can be used remotely, if the attacker has system level privileges (or root in Unix-based systems).

Another way to destabilise hardware is to alter the software that controls communications with it. Altering "driver" files prevents the device from being used until the driver is manually reinstalled. For example, changing the settings of a video adaptor to disable the displays disrupts a hard disk array by changing its composition. Though currently, such changes are becoming harder to perform due to increased levels of protection deployed at the operating system level.

Changing the algorithm or altering the transmitted data of a single unit in line can result in a chain reaction malfunction for multiple mechanical units. Access to the ICS configurations can result in physical damage to the equipment. Altering the displayed data and status information can result in mechanical components working in an abnormal state, while human-machine interface or SCADA indicate that all the values are nominal, damaging the components and systems. Once the system is damaged, it is not always possible to identify, if the source of damage was a remote cyberattack or a local mechanical malfunction.

The system level privileges on the target computer give the attackers an unlimited control over the system. To prevent attackers from gaining system-level rights, defence should involve system enhancement, regular software updates, and personnel awareness on the consequences of abusing computer systems.

Exfiltration Tools

Exfiltration tools are used to collect and extract valuable data from the target system. Most common techniques allow us to:

- collect system credentials and configurations, locate valuable files;
- transfer files via simple transfer scripts and minimal web servers (if the security level is low);
- use encryption or steganography to disguise the data and transfer it to an intermediate server or cloud;
- avoid blocked ports and reroute data through the common protocols that are normally allowed by the network policies (e.g., HTTPS, SSH, FTP, DNS);
- reconfigure firewalls, transfer data, and then return the original configurations back.

Steganographic programs and scripts allow for the creation of files which contain a certain amount of random noise or legitimate information, such as graphic, video, or audio files, that can be further used to embed hidden data (small files) without affecting the inconspicuous content. Thus, valuable data can be safely exfiltrated by placing an image object on an externally facing website (website that can be accessed from the public network) in a background graphic or logo attached to an email, or even an audio message.

Even in highly secure environments, there is a probability of popular outgoing protocols (SSH, HTTP/HTTPS, mail protocols, etc.) to be "allowed" or "filtered" by firewalls and network configurations, but not "blocked". Infrastructure related protocols, such as DNS and DHCP, are allowed and unmonitored as they cannot be exploited in a conventional way. Those unrestricted protocols can be used by attackers as covert channels for malware control or data exfiltration. In the case of mail protocols, it is possible to transfer information via encoded and truncated form (split into multiple network packets).

Attackers can create a tunnel to forward Secure Shell (SSH) over HTTP/HTTPS and utilise the provided proxy server to exfiltrate the data. As SSH connection provides full interactive command line (shell), the intruders can simply copy the files with no additional tools.

The defence solution is to deploy and configure advanced defence software and network security policies, as well as to raise the awareness of staff.

Persistence Tools

To avoid potential detection, it is not recommended to use the same exploit to access the network multiple times. Once the network is breached, the attackers have to ensure the continuous access to the system by adding either "Authorized" access through creating additional user accounts, or adding backdoors to a system or application. One of the most well-known backdoors can be created using the *"netcat"* tool, but the attackers may also covertly install remote administration tools, VPN servers, FTP and web servers, to open additional network channels of entry.

To mitigate the effect of persistence tools both outgoing and incoming traffic must be filtered while preserving sufficient functionality of the system and the network. In addition, the routine audit of accounts, system access, open ports, and other items that could be used to create a backdoor, should be performed. It is possible to lock administrative access to internal systems using additional utilities. These tools will help to prevent the deployment of backdoors and make it significantly harder for those attackers who were able to obtain even the highest privilege level in the system.

In the virtualised networks, it is possible to reset the employee workstations to the default non-infected state (provided by a preconfigured virtual image) every few days or weeks, wiping any potential persistent malware installed in the system. In this case all employee data is stored in the remote server memory, thus making the server or a group of servers the only critical asset for the organization. However, if the original virtual image is infected with the malware, every computer in the virtual network becomes infected after a network-wide reset (restoration from the virtual image).

Obfuscation Tools and Covering Tracks

In cyber security "obfuscation" means concealing the presence while operating in the system or network. The key tasks in stealth operations are: hiding original location, hiding attack operating system features and tools, editing event logs, and manipulating configuration files.

To hide or obfuscate the original location, the attackers use remote computers (Proxy server, VPN server, Tor, etc.). Those intermediate nodes can be purposely configured, rented, or deployed on compromised computers, IoT devices, or network routers through which the intruders are operating.

Proxy networks like Tor, Bitblinder, Perfect Dark, I2P, Freenet allow the attackers, of both beginner and expert level, to use simple and automated tools to hide their original location. Those systems use a different underlying protocol (Tor uses TCP, I2P uses UDP, etc.) providing various levels of anonymity, confidentiality and integrity, as well as affecting the bandwidth speed significantly due to additional layers of encryption and slow communication speeds between nodes. Additionally, to ensure an extra layer of anonymity, Virtual Private Networks (VPNs) can be used.

Actions and manipulations with the computer system generate entries in the event history records (logs) of systems and network devices. At the end of every connection the attackers remove such traces to hide their presence from potential investigators, system administrators and forensic analysts.

In the Unix-based systems with sufficient privileges, under certain conditions event history files (logs) can be changed using a simple text editor. Thus, in secure server systems those files cannot be deleted or edited, as they have "append only" permissions. To circumvent this defence measure and to cover their tracks, the attackers might either disable logs completely in the very beginning of the operation, generate large amounts of false events during or after the operation, or cause a Permanent Denial of Service at the end of the operation (destroying the equipment).

Kernel-level malware (rootkits) can be used to hide files and isolate via the system level directives nearly any tool or utility from finding malicious agents.

To combat stealth techniques, honeypots and honeynets are designed and deployed. They create fake virtual computers and networks to lure the attackers and make them generate additional network activity for further forensic analysis. The intrusion prevention system can also potentially force the attacked system to communicate with the attacker, to gather additional information.

In the case of file and log manipulation, tools, such as an Open Source Tripwire, can be used to monitor file manipulations in real-time and issue alerts when specific changes are made or specific files are accessed. Additionally, the server can be configured to send copies of the log entries to remote network access storage. These two measures will ensure that if such attacks occur, the activity will be recorded in multiple places in the network.

Cyber tools *per se* can surpass the limitations of cyberspace and affect the physical world by compromising cyber-physical systems.

1.1.3 Cyber-physical Systems

Cyber-physical systems are physical platforms, partly or fully autonomous, that are controlled by computerised control systems. They include, but are not limited to, IoT devices, ICS, vehicles, vessels, aircrafts, robots, and autonomous weapon systems (AWS).

Automated command and control systems, which were invented more than half a century ago, are now equipped with powerful computers and are managing and guiding military devices, air, naval and ground AWS, including combat robots, remotely-piloted systems (aircrafts, drones), cruise missiles, submarine-hunting robot ships, and robot tank armies. The next generation weapon systems, remotely controlled, semi-autonomous and autonomous, are being developed, tested, and deployed all over the world.

Fast developing and AI-enhanced programs and control systems with sophisticated logic and computer vision software enabled weapons implement intelligent functions. They perform advanced data analysis and processing, including data mining (e.g., to analyse geolocation data for the potential locations of nuclear missiles[62]), complex decision making, problem solving, forecasting, pattern recognition, intrusion detection, systems targeting[63], and autonomous "jam-proof" navigation[64].

> Robotics and Autonomous Weapon Systems are gaining their place in our daily life, as well as in military operations. As per the forecasts of a UK expert, the US military will have more robot soldiers on the battlefield than real ones by 2025. A GCHQ security consultant believes deadly combat robots are rapidly becoming "a reality" of modern day warfare.[65]

Cyber-physical weapons have reached a highly sophisticated level of automation, that under certain conditions allows them full autonomy. Autonomous weapons are considered to be the Fourth Generation Warfare (4GW), after gunpowder (second generation) and nuclear arms (third generation). They are defined as "a weapon system that, once activated, can select and engage targets without further intervention by a human operator. This includes human-supervised autonomous weapon systems that are designed to allow human operators to override the operation of the weapon system but can select and engage targets without further human input after activation."[66]

> **The US Army Robotics and Autonomous Systems strategy is aimed to:**
> - Sustain current systems.
> - Maintain the current fleet of tele-operated UGSs and remotely piloted UAS.
> - Recapitalize older robots.
> - Improve existing systems.
> - Field a universal controller for legacy and new programs.
> - Field autonomous technologies within UGS and UAS where possible.
> - Refine automated ground resupply operations as the Army's first semi-autonomous vehicle.

---

[62] P. Stewart, Deep in the Pentagon, a secret AI program to find hidden nuclear missiles, 5 June 2018, last retrieved on 19 July 2018 at https://www.reuters.com/article/us-usa-pentagon-missiles-ai-insight/deep-in-the-pentagon-a-secret-ai-program-to-find-hidden-nuclear-missiles-idUSKCN1JJ114J

[63] W. Webb, US Army Developing Drones with AI targeting, 23 April 2018, last retrieved on 29 May 2018 at http://truthinmedia.com/us-army-developing-drones-artificial-intelligence-ai-targeting/

[64] M. Peck, Army wants smarter computer AI for electronic warfare, 2 August 2017, last retrieved on 14 July 2018 at https://defensesystems.com/articles/2017/08/02/army-electronic-warfare-ai.aspx

[65] J. Locketi, US military will have more combat robots than human soldiers by 2025, 15 June 2017, last retrieved on 8 December 2018 at https://nypost.com/2017/06/15/us-military-will-have-more-combat-robots-than-human-soldiers-by-2025/

[66] M. Gubrud, DoD Directive on Autonomy in Weapon Systems, 27 November 2012, last retrieved on 16 July 2018 at https://www.icrac.net/dod-directive-on-autonomy-in-weapon-systems/

Develop new capabilities
- Develop off-road autonomy for unmanned combat vehicles.
- Develop swarming for advanced reconnaissance.
- Develop artificially intelligent augmented networks and systems.
- Replace obsolete systems.
- Replace non-standard equipment systems with new programs of record.
- Replace manned systems with unmanned systems to allow soldiers to perform other tasks.

Assess new technologies and systems
- Continue assessments on the state of UGS and UAS autonomy to ensure systems progress with available technology.
- Determine where technologies can serve cross-domain solutions, especially with new payloads.

Source: The US Army Robotics and Autonomous Systems strategy[67], March 2017

AWS demonstrate increased speed, surpass human limitations, save costs and spare human personnel from risks in dangerous zones. The US Army Robotics and Autonomous Systems strategy[68], published in 2017, specifically references the ability of autonomous systems to increase effectiveness at a lower cost and risk. They are cheaper to maintain; they work with 100 percent efficiency (with zero distractions and downtime) and can be easily repaired or replaced.

According to the level of human intervention in the process of selecting and attacking targets, AWS can be sub-divided into:[69]

- Human-in-the-Loop Weapons: robots that can select targets and deliver force only with a human command.
- Human-on-the-Loop Weapons: robots that can select targets and deliver force under the oversight of a human operator who can override the robots' actions.
- Human-out-of-the-Loop Weapons: robots that are capable of selecting targets and delivering force without any human input or interaction.

Emerging danger, which is now discussed globally, represents the Lethal Autonomous Weapon Systems (LAWS, also called "killer robots") – a type of autonomous military robots that can apply lethal force; systems that are capable of taking decisions autonomously, select and attack targets without a human operator.

---

[67] The US Army Robotics and Autonomous Systems strategy, March 2017, last retrieved on 3 June 2018 at http://www.arcic.army.mil/App_Documents/RAS_Strategy.pdf

[68] Ibid

[69] Human Rights Watch, Expert meeting, Autonomous Weapon Systems, Technical, Military, Legal and humanitarian aspects, Geneva, 2014, B. Docherty, Losing Humanity: The Case Against Killer Robots, Human Rights Watch, November 2012, p. 2,http://www.hrw.org/sites/default/files/reports/arms1112_ForUpload.pdf

The UN Report of the Special Rapporteur[70] on extrajudicial, summary or arbitrary executions, Christof Heyns, 2013, states: "While much of their development is shrouded in secrecy, robots with full lethal autonomy have not yet been deployed. However, robotic systems with various degrees of autonomy and lethality are currently in use, including the following:

- The US Phalanx system for Aegis-class cruisers automatically detects, tracks and engages anti-air warfare threats such as anti-ship missiles and aircraft.
- The US Counter Rocket, Artillery and Mortar (C-RAM) system can automatically destroy incoming artillery, rockets and mortar rounds.
- Israel's Harpy is a 'Fire-and-Forget' autonomous weapon system designed to detect, attack and destroy radar emitters.
- The United Kingdom Taranis jet-propelled combat drone prototype can autonomously search, identify and locate enemies but can only engage with a target when authorized by mission command. It can also defend itself against enemy aircraft.
- The Northrop Grumman X-47B is a fighter-size drone prototype commissioned by the US Navy to demonstrate autonomous launch and landing capability on aircraft carriers and navigate autonomously.
- The Samsung Techwin surveillance and security guard robots, deployed in the demilitarized zone between North and South Korea, detect targets through infrared sensors. They are currently operated by humans but have an 'automatic mode'.

The 2016 publication "Killer Robots: Lethal Autonomous Weapon Systems Legal, Ethical and Moral Challenges", claims that "though LAWS in the true sense are not yet available, at least 44 countries, including China, France, Germany, India, Israel, South Korea, Russia, the UK and the United States, are developing such capabilities." [71]

> "The real leading military robotics developers in the world are – in no particular order – the US, China, Russia, Israel, the UK, France, and South Korea; all doing really pretty interesting and sophisticated things, and all approaching this question of delegating lethal force a little bit differently. Pentagon leaders have been very clear that their intention is to keep humans in charge. They've acknowledged that they might shift if other countries cross that line, that they might have to as a result." [72]
>
> *Paul Scharre, Senior fellow and Director, Technology and National Security Program Center for a New American Security (CNAS)*

---

[70] UN Report of the Special Rapporteur on extrajudicial, summary or arbitrary executions, Christof Heyns, 9 April 2013, last retrieved on 3 July 2018, https://www.ohchr.org/Documents/HRBodies/HRCouncil/RegularSession/Session23/A-HRC-23-47_en.pdf

[71] Wing Commander (Retd) Dr U C Jha, Killer Robots: Lethal Autonomous Weapon Systems Legal, Ethical and Moral Challenges, 2016, Vij Books, New Delhi, India.

[72] Ibid

What LAWS Are Not
Lethal autonomous weapons should not be confused with unmanned combat aerial vehicles or "combat drones", which are currently remote-controlled by human pilots. They fly but do not fire autonomously.
Weaponised combat suits and exoskeletons cannot be considered LAWS either, as they are solely controlled by an operator, and only aid in performing tasks, without any autonomous decision making involved.

As per the "Military Robots Market – Global Forecast to 2022" report[73] (2017), the military robotics market is projected to grow from 16.79 billion USD in 2017 to 30.83 billion USD by 2022, at a CAGR[74] of 12.92 percent during the forecast period. Factors such as focusing on the development of unmanned military systems, use of robots for a new range of military applications, and military modernization programs in various countries are expected to drive the growth of the military robotics market from 2017 to 2022.[75]

The AI in military market is projected to grow from 6.26 Billion USD in 2017 to 18.82 Billion USD by 2025, at a CAGR of 14.75% during the forecast period.[76] Military forces worldwide are focused on the integration of AI with unmanned weapon systems, such as automated robots, self-driving military vehicles, and UAV control systems. The absence of standards and protocols for the use of AI in military applications is acting as a key restraint to the growth of the artificial intelligence in the military market.

Research and Markets 2018 Report "Artificial Intelligence in Military Market by Offering (Software, Hardware, Services), Technology (Learning & Intelligence, Advanced Computing, AI Systems), Application, Platform, Region - Global Forecast to 2025"

In the "Army of None: Autonomous Weapons and the Future of War", Paul Scharre concludes: "There are at least 90 countries that have drones today, and 16 countries are counting that they have armed drones, including many non-state groups. A lot of them are remotely controlled or tele-operated... There are at least 30 countries that have these sort of automatic mode(s) - it's kind of wartime modes – that they can turn on these systems that are either land-based, air-and-missile defence systems,

---

[73] Military Robots Market by Platform, Application, Mode of Operation, and Region-Global Forecast to 2022, Report, September 2017, last retrieved on 1 October 2018 at https://www.researchandmarkets.com/research/8s436m/military_robots
[74] CAGR – compound annual growth rate
[75] Military Robots Market by Platform, Application, Mode of Operation, and Region-Global Forecast to 2022, Report, September 2017, last retrieved on 1 October 2018 at https://www.researchandmarkets.com/research/8s436m/military_robots
[76] Artificial Intelligence in Military Market by Offering, Technology, Application, Platform, Region – Global Forecast to 2025, Report, March 2018, last retrieved on 1 October 2018 at https://www.researchandmarkets.com/research/kktnpz/artificial?w=4

or they're on ground vehicles or ships, that allow this automatic protection bubble to kick in."[77]

Built with advanced technologies, cyber-physical systems are not immune to cyberattacks. The embedded devices (IoT, ICS, SCADA, etc.) and robotics (UAVs, UGVs, etc.) represent a physical extension of the cyber domain, and can be used as additional attack vectors, access points and platforms to launch cyber attacks. Furthermore, AI-enhanced systems are becoming increasingly vulnerable to AI specific attacks.

The key vulnerable electronically controlled mechanical components of the above-mentioned systems, include transportation systems (wheels, threads, legs, flight engines, etc.), targeting modules, and secondary mechanical components (vehicle subsystems). The consequences of these cyber vulnerabilities are particularly severe for autonomous systems that are used for high-risk activities.

Evolving autonomy of weapons is changing the military environment and concepts. A single human operator can now control an army of autonomous systems (e.g., drones, computer viruses, etc.). Moreover, the software components for such attacks are increasingly available even for civilian users (e.g., open source face recognition, navigation, planning algorithms, multi-agent swarming frameworks).

Highlights

2000–2004

In 2000, the US Congress ordered that one-third of military ground vehicles and deep-strike aircraft should be replaced by robotic vehicles. Six years later, hundreds of PackBot Tactical Mobile Robots were deployed in Iraq and Afghanistan to open doors in urban combat, lay optical fibre, defuse bombs and perform other hazardous duties that would have otherwise been carried out by humans.[78]

The Talon robot was initially deployed for military operations in Bosnia in 2000 and has been in service with the US military since 2001. It can be deployed in military, first responder and law enforcement applications, and can be reconfigured to conduct a range of missions, including CBRNe/hazmat, explosive ordnance disposal, rescue, heavy lift, communications, security, reconnaissance and detection of mines, unexploded ordinance and improvised explosive devices (IEDs). It also

---

[77] P. Scharre, Killer Robots and Autonomous Weapons With Paul Scharre, podcast, 1 June 2018, last retrieved on 2 August 2018 at https://www.cfr.org/podcasts/killer-robots-and-autonomous-weapons-paul-scharre

[78] S. Parkin, Killer robots: The soldiers that never sleep, 16 July 2015, last retrieved on 3 July 2018 at http://www.bbc.com/future/story/20150715-killer-robots-the-soldiers-that-never-sleep

supports special weapons and tactics, and military police operations. This robot was used at Ground Zero during the recovery mission of 9/11.[79]

Further, the SWORDS robot was designed, capable of carrying a range of weapons for special weapons observation reconnaissance detection system (SWORDS) duties. Additional weapons, such as a grenade launcher, rocket launcher, or machine guns can be installed onto the robot. It can act in any hazardous conditions, on any terrain and even underwater.

As early as 2005 the New York Times reported the Pentagon's plans to replace soldiers with autonomous robots.[80] The further evolutionary step was supposed to include smart devices able to make their own decisions.

2008

The first Modular Advanced Armed Robotic System (MAARS) is delivered. It is an unmanned ground vehicle designed for reconnaissance, surveillance and target acquisition as well as Force Protection missions to increase the security of personnel manning forward locations. MAARS can be positioned in remote areas where personnel are currently unable to monitor their security. Utilizing the suite of security sensors on MAARS, the small unit can achieve early warning of potential threats and immediately engage, if necessary. MAARS can be operated from over 800 metres away using either the wearable, Tactical Robotic Controller or a Toughbook Laptop Controller. Both controllers allow for the piping of a video feed to a TV for observation in a Tactical Operations Center.[81]

2010

The South Korean Super aEgis II automated turret, unveiled in 2010, is used both in South Korea and in the Middle East. It can identify, track, and destroy a moving target at a range of 4 km, theoretically without human intervention. The turret is currently in active use in numerous locations in the Middle East, including three airbases in the United Arab Emirates (Al Dhafra, Al Safran and Al Minad), the Royal Palace in Abu Dhabi, an armoury in Qatar and numerous other unspecified airports, power plants, pipelines and military airbases elsewhere in the world.[82]

---

[79] TALON Tracked Military Robot, last retrieved on 3 July 2018 at https://www.army-technology.com/projects/talon-tracked-military-robot/

[80] S. Parkin, Killer robots: The soldiers that never sleep, 16 July 2015, last retrieved on 3 July 2018 at http://www.bbc.com/future/story/20150715-killer-robots-the-soldiers-that-never-sleep

[81] Modular Advanced Armed Robot System (MAARS), last retrieved on 3 October 2018 at https://www.qinetiq-na.com/wp-content/uploads/brochure_maars.pdf

[82] S. Parkin, Killer robots: The soldiers that never sleep, 16 July 2015, last retrieved on 3 July 2018 at http://www.bbc.com/future/story/20150715-killer-robots-the-soldiers-that-never-sleep

2016

Russia unveiled a fully driverless armed ground vehicle Uran-9, a 10-ton system armed with anti-tank missiles (9M120-1 Ataka), rockets (Shmel-M), reactive flamethrowers, 2A72 automatic cannon and a 7.62mm coaxial machine gun. The Uran-9 is controlled by an operator in a mobile vehicle (no more than 2.9 km away) who can either manually control it or set it on a pre-programmed path. It is equipped with a variety of sensors, laser warning systems, thermal and electro-optic cameras.[83]

The Titan UGV was unveiled at the 2016 Association of the United States Army Annual Meeting. It was also displayed at the ninth annual Ground Vehicle Systems Engineering and Technology Symposium. Titan is a multi-mission UGV that can be reconfigured to enhance mission effectiveness. It integrates the battlefield-tested robotic systems and controller from the QNA and THeMIS UGV platform, and modular mission payload developed by Milrem. The platform includes two track modules connected by a payload frame. The open architecture of the Titan UGV allows for the integration of mission equipment and systems to support a range of operations.[84]

2017

Russian weapons manufacturer Kalashnikov Group developed a range of combat robots that are fully automated and use AI to identify targets and make independent decisions. The Group also announced the range of products based on neural networks, including a fully automated combat module featuring the same technology.[85]

Israel's Meteor Aerospace Ltd. unveiled RAMBOW, its latest unmanned ground vehicle, at the Autonomous Unmanned Systems and Robotics 2017 Air Show in Rishon Lezion. The RAMBOW can drive autonomously along predefined routes, automatically detecting and avoiding obstacles using numerous on-board sensors, at a maximum speed of 50 km/h.[86]

---

[83] Russia says it has deployed its Uran-9 robotic tank to Syria — here's what it can do, 15 March 2018, last retrieved on 13 January 2019 at https://www.businessinsider.com/russia-uran-9-robot-tank-what-can-it-do-syria-2018-5?r=UK&IR=T

[84] Titan Unmanned Ground Vehicle (UGV), last retrieved on 3 October 2018 at https://www.army-technology.com/projects/titan-unmanned-ground-vehicle-ugv/

[85] R. Haridy, Kalashnikov's new autonomous weapons and the "Terminator conundrum", 21 July 2017, last retrieved on 15 October at https://newatlas.com/kalashnikov-ai-weapon-terminator-conundrum/50576/

[86] RAMBOW: Israel unveils latest unmanned ground vehicle, 18 September 2017, last retrieved on 13 January 2019 at https://www.jpost.com/Israel-News/Meet-RAMBOW-the-latest-unmanned-ground-vehicle-505409

2018

United Kingdom

From Wednesday night to Friday morning, from the 19th to 21st December 2018, flights in and out of Gatwick were halted after a small drone, or perhaps multiple drones, were spotted over the airfield.[87] British authorities deployed a range of measures, including helicopters, police officers on the ground, and a military-grade system to find the drone or drones. About 1,000 aircrafts were either cancelled or diverted, affecting about 140,000 passengers, during three days of disruption.[88] Shortly after this event, on 8 January 2019, London's Heathrow Airport briefly halted departures after the report of a drone sighting.[89]

Why airports can't stop drones from causing chaos

"It seems absurd that a couple of $500 drones and some rogue pilots can disrupt millions of dollars of business by causing flight delays and negative experiences for travelers."

Brian Wynne, president and CEO of the Association for Unmanned Vehicle Systems International, wants to see the drone industry and aviation authorities around the world agree to 'remote identification,' something like digital license plates, for drones. With remote identification, he believes, airports could easily identify a drone's owner and contact them to land the drone if it's flying where it shouldn't.

The drone sightings at Heathrow and Gatwick prompted quick regulatory changes. Law enforcement in the U.K. is now permitted to land, seize and search drones, and to fine operators who fail to comply with orders to land their small drones, or fine those who fly without properly registering their drones as well."

CNBS[90]

Syria

A swarm of armed miniature drones attacked the main Russian military base in Syria. More than a dozen armed drones descended from an unknown location, allegedly from between 50 and 100 km away, onto Russia's vast Hmeimim air base in north-western Latakia province, and on the nearby Russian naval base at Tartus. Russia said that it shot down

---

[87]   B. Stansall, How To Stop A Drone? There's No Good Answer, 21 December 2018, last retrieved on 23 December 2018 at https://www.npr.org/2018/12/21/679293917/how-to-stop-a-drone-theres-no-good-answer

[88]   Gatwick drone: Labour calls for independent inquiry, 23 December 2013, last retrieved on 24 December 2018 at https://www.bbc.co.uk/news/uk-politics-46663898

[89]   L. Josephs, London's Heathrow airport briefly halts flights after 'drone sighting', 8 January 2019, last retrieved on 14 January 2019 at https://www.cnbc.com/2019/01/08/londons-heathrow-airport-halts-departures-after-drone-sighting.html

[90]   L. Kolodny, Why airports can't stop drones from causing chaos, 9 January 2019, last retrieved on 14 January 2019 at https://www.cnbc.com/2019/01/08/why-airports-cant-stop-drone-disruptions.html

seven of the 13 drones and used electronic countermeasures to safely bring down the other six.[91]

China

In May, China's Ehang broke the world record for the largest number of drones by deploying a fleet of 1,374 drones at a Labour Day show in the city of Xi'an to beat the 1,218 drones flown by Intel in February at the Pyeongchang 2018 Winter Olympic Games. The drone maker was paid 10.5 million yuan (1.6 million USD) to perform its record-breaking feat at the Labour Day show, according to an announcement posted on the official platform for government procurement information run by the Ministry of Finance. The show's live broadcast footage, however, showed about half of the drones in flight were in disarray. The Global Positioning System mechanisms on 496 of the drones in the show were interfered with, according to a statement on the company's Weibo account.[92]

Israel

The Harpy anti-radar "fire and forget" drone is designed to be launched by ground troops, and autonomously fly over an area to find and destroy radar that fits pre-determined criteria. Israel sold fully autonomous drones capable of attacking radar installations to China, Chile, India, and other countries. In 2017, Israeli Minister Ayoub Kara stated that Israel is developing military robots as small as flies in order to assassinate leaders of Hezbollah and Hamas, and that these robots may be operational within as few as three years.[93]

Guardium, developed by G-NIUS, is an Israeli unmanned ground vehicle (UGV) used to combat and guard against invaders. It can be used either remotely or in autonomous mode. Both modes do not require human interaction for the vehicle to work; thus, giving the operator a full audio and visual view of the vehicle's surroundings at all times. When working in groups these vehicles can work together and cooperate within the network.[94]

---

[91] L. Sly, Who is attacking Russia's bases in Syria? A new mystery emerges in the war, 9 January 2018, last retrieved on 15 October at https://www.washingtonpost.com/world/who-is-attacking-russias-main-base-in-syria-a-new-mystery-emerges-in-the-war/2018/01/09/4fdaea70-f48d-11e7-9af7-a50bc3300042_story.html?noredirect=on&utm_term=.d0e077e8f87e

[92] I. Deng, China's Ehang broke the world drone display record – but the aerial bots were out of sync, 2 May 2018, last retrieved on 2 August 2018 at https://www.scmp.com/tech/enterprises/article/2144393/chinese-start-broke-world-drone-display-record-aerial-bots-werent

[93] J. Daniels, Mini-nukes and mosquito-like robot weapons being primed for future warfare, 17 March 2017, last retrieved on 15 October 2018 at https://www.cnbc.com/2017/03/17/mini-nukes-and-inspect-bot-weapons-being-primed-for-future-warfare.html

[94] A. May, "Phantom on the fence". Israel Defense Forces, last retrieved on 10 January 2019 at https://www.israeldefense.co.il/en/company/g-nius-unmanned-ground-systems

Russia

The Kalashnikov Group tested a new Unmanned Ground Combat Vehicle, the first robotic fighting vehicle, called Soratnik (Comrade-in-Arms) under "near-combat" conditions. The system is designed for reconnaissance, fire support, patrols and for the protection of areas and important facilities, as well as mine clearance and obstacle clearing. This is an armoured tracked vehicle equipped with a computer control system. In addition, a Remote-Controlled Weapon Station can be installed, and Soratnik can operate in conjunction with other automated combat units, including the ZALA AERO, unmanned aircraft.[95]

USA

The Defense Department proposed a 25 percent increase over 2018 funding for the FY2019 for unmanned systems and associated technologies, which would allow it to reach 9.39 billion USD. That proposal includes funding for 3,447 new air, ground, and sea drones. In 2018, the budget called for only 807 new drones, according to the report.[96]

The US has developed and deployed many military AI combat programs, such as the Sea Hunter autonomous warship, which is designed to operate for extended periods at sea without a single crew member, and to even guide itself in and out of port.

BigDog - the First Advanced Rough-Terrain Robot

BigDog, being the size of a large dog or small mule, has four legs that mimic an animal with compliant elements to absorb shock and recycle energy from one step to the next. The on-board computer controls locomotion, processes sensors, and handles communications with the user. The control system keeps the robot balanced, manages locomotion on a wide variety of terrain, and does navigation. Sensors for locomotion include joint position, joint force, ground contact, ground load, a gyroscope, LIDAR, and a stereo vision system. Other sensors focus on the internal state of BigDog, monitoring the hydraulic pressure, oil temperature, engine functions, battery charge, and others. BigDog runs at 10 km/h, climbs slopes up to 35 degrees, walks across rubble, climbs muddy hiking trails, walks in snow and water, and carries up to 150 kg loads. Development of the original BigDog robot was funded by DARPA. Further work to add a manipulator and do dynamic manipulation was funded by the Army Research Laboratory's RCTA program.[97]

---

95 BAS-01G Soratnik "Comrade in Arms", last retrieved on 17 October 2018 at https://www.globalsecurity.org/military/world/russia/soratnik.htm

96 D. Gettinger, Study: Drones in the FY 2019 Defense Budget, 9 April 2018, last retrieved on 18 October 2018 at http://dronecenter.bard.edu/drones-in-the-fy19-defense-budget/

97 BigDog - The First Advanced Rough-Terrain Robot, last retrieved on 7 November 2018 at https://www.bostondynamics.com/bigdog

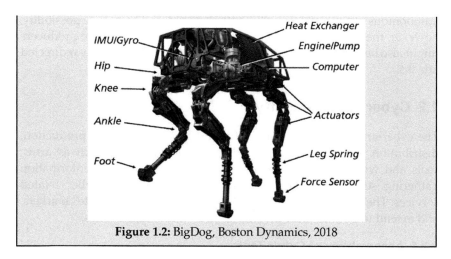

Figure 1.2: BigDog, Boston Dynamics, 2018

In 2018, the US Army's Universal Ground Control Station was enhanced by AI after the military contracted with Stryke Industries and their sub-contractor Scorpion Computer Services to support drone operations and provide the highest level of cybersecurity. The operators will be able to control even groups of drones, e.g., MQ-1C Gray Eagle, following the application of Scorpion's AI system Scenario Generator (ScenGen).[98]

The unique ScenGen system can test internal systems, automating regression, and plan military operations. Its speed of thinking equals to 250 years of human work every 90 minutes, it protects every potential point of entry against cyberattacks and reacts immediately. However, the CEO of Scorpion Computer Services forecasts that with further technological development ScenGen might be hackable by AI quantum computing systems, thus it is essential to make regular updates in order to be ready for the next generation of cyberwar when the winner is the one who is the best at maths."[99]

Research is ongoing in development of autonomous systems in air, ground, on water and underwater vehicles. Modern cyber-physical systems have many challenges, including the unpredictable behaviour of the AI-enhanced software, high research and development cost, deployment and use in an unclear legal environment.

Autonomous AI enhanced weapons are changing the war interface, contributing to the military capabilities of countries and increasing the scope of deterrence. The worst-case scenario is the battle of AI enhanced

---

[98] M.R. Bauer, The next cyber arms race is in artificial intelligence, January 2018, last retrieved on 8 December 2018, at https://www.fifthdomain.com/dod/2018/01/24/the-next-cyber-arms-race-is-in-artificial-intelligence/

[99] Ibid

autonomous weapons from all conflicting sides. There is a possibility, however, that after the release of such intelligent weapons, the producers might also become the receivers of the intelligently and covertly redirected attacks.

1.2 Cyber-Arms Industry

The cyber-arms industry includes the design, development, production, distribution and acquisition of cyber related products, such as arms, tools and weapons (e.g., exploits, zero-day vulnerabilities, information gathering scripts, malware, etc.) and the provision of cyber related services. The distribution and deals take place at official "white" markets, and extend to both online and offline "grey" and "black" ones.

1.2.1 Accessibility of Cyber Tools

The duality of cyber tools allows the use of the same tools for both defensive and offensive operations. With the overall automation of virtual communication and evolution of computer business, commercially available cyber tools are becoming more accessible to the nonprofessional public.

The majority of cyber tools are free and available for download as open source. There are thousands of tutorials on YouTube and other social media platforms teaching novices how to implement malicious attacks, even with a catchy name "Start hacking today". Thus, even the beginner level users can use publicly available tools, activated by simply inputting the IP address and pressing a "Fire" button (e.g., LOIC Denial of Service tool). These activities are easy to perform, and easy to detect. The case is more complicated when more sophisticated hackers easily acquire any necessary tool either at a low price, for free through various internet leaks, or through the highest bidder at the black market.

In March 2017, WikiLeaks wrote:

Recently, the CIA lost control of much of its hacking arsenal, including malware, viruses, trojans, weaponized "zero day" exploits, malware remote control systems and associated documentation. This extraordinary collection, which amounts to more than several hundred million lines of code, gives its possessor the entire hacking capacity of the CIA. The archive appears to have been circulated among the former US government hackers and contractors in an unauthorized manner, one of whom has provided WikiLeaks with portions of the archive. "Year Zero" introduces the scope and direction of the CIA's global covert hacking program, its malware arsenal and numerous exploits weaponised with "zero day" vulnerabilities against a wide range of the US and European consumer products, including Apple's iPhone, Google's Android and Microsoft's Windows and even Samsung TVs, which are turned into covert microphones.[100]

Julian Assange, the editor of WikiLeaks, stated that "there is an extreme proliferation risk in the development of cyber 'weapons'. Comparisons can be drawn between the uncontrolled proliferation of such 'weapons', which result from the inability to contain them combined with their high market value, and the global arms trade. But the significance of "Year Zero" goes well beyond the choice between cyberwar and cyberpeace. The disclosure is also exceptional from a political, legal and forensic perspective."[101]

"Year Zero", the first full part of the "Vault 7"series of leaks, comprised 8,761 documents and files from an isolated, high-security network situated inside the CIA's Center for Cyber Intelligence in Langley, Virginia. It followed an introductory disclosure of the CIA targeting French political parties and candidates in the lead up to the 2012 presidential election.[102]

Not only cyber tools, but also pre-trained neural networks can be also downloaded for free, and it is possible to buy a drone, train it for face recognition and use it as a weapon platform. Without any expert knowledge, these drones can be programmed further and used for identifying objects, including people.

Custom-built racing drones (that can fly at the speed up to 100 km/h) can be made with a budget of around 250 USD. Alternatively, an already assembled and ready-to-fly drone costs in average 5 to 10 times more, and yet it is still affordable for advanced operations by amateurs. It usually comes with technological building blocks that are available to anyone – video cameras, open-source software and low-cost computer chips.

There are claims that Moore's law has changed.[103] Instead of new chips coming out every 18 months at twice the speed but at the same cost as their predecessors, new chips are coming out at the same speed as their predecessors but at half the cost. In the historical perspective, computers used in the Apollo Space Program were ten years ahead of their time, yet the hardware of a single modern smartphone can easily replace all the computers used by the Program.[104] For modern cyber arms the life-span could be weeks depending on the timeline for discovering the same vulnerability or patching it.

---

[100] Vault 7: CIA Hacking Tools Revealed, Press release, 7 March 2017, last retrieved on 14 December at https://wikileaks.org/ciav7p1/

[101] Ibid

[102] WikiLeaks drops 'Grasshopper' documents, part four of its CIA Vault 7 files, Wired, 7 May 2017, last retrieved on 11 January 2019 on https://www.wired.co.uk/article/cia-files-wikileaks-vault-7

[103] a16z Podcast: Software Programs the World, with Marc Andreessen, Ben Horowitz, Scott Kupor, and Sonal Chokshi, last retrieved on 14 December 2018 at https://a16z.com/2016/07/10/software-programs-the-world/

[104] D. Grossman, How Do NASA's Apollo Computers Stack Up to an iPhone?, 13 May 2017, last retrieved on 14 December 2018 at https://www.popularmechanics.com/space/moon-mars/a25655/nasa-computer-iphone-comparison/

As a countermeasure to the above, there exist a number of solutions which could decrease cyber tools accessibility and increase control through:

- increasing the cost of the security assessment and audit tools
- licencing any computer related to cyber security or, at least to, cyber security related activities
- licencing of products and the registration of sold equipment.

Cyber experts and communities voice the necessity to administer cyber certifications and develop criteria for evaluating the existing certification programs. There should be a national, and even an international, body to register and license cyber professionals and authorize cyber certifications. The mandatory personal signature on the used tools could also allow for the trace of the chain of users.

One of the best global practices in education and certification is demonstrated by the activities of the European Computer Driving Licence Foundation (ECDL) – an international organization dedicated to raising digital competence standards in the workforce, education and society, and providing computer certification. ECDL has grown from a small project in Europe, founded with support from the European Union's ESPRIT research programme, to an internationally recognised standard in digital skills. Its certification programmes, available in over 40 languages, networks in more than 100 countries, enables individuals and organizations to assess, build and certify their competence in the use of computers and digital tools to the globally-recognised ECDL standard, known as International Computer Driving Licence in non-European countries.[105]

In 2015, a technical intelligence wing of the Indian government, National Technical Research Organization proposed the creation of the Indian National Cyber Registry that would act as a platform for directly connecting IT professionals with various government agencies. "... I hereby propose creation of an Indian National Cyber Registry. The INCR will be a national repository of IT professionals," its Chairman Alok Joshi said at a cybersecurity event Ground Zero Summit 2015.[106]

Though information technology and information security are rapidly changing and evolving disciplines in a relatively new field, their products should pass the same licencing procedure as in medicine, aviation, etc. The development and adoption of standard international regulations on the use of cyber arms by non-professionals will help to avoid many of unexpected threats and damaging consequences, including cybercrimes.

---

[105] About ECDL Foundation, ECDL Foundation, last retrieved on 14 October 2018 at http://ecdl.org/about

[106] NTRO proposes creation of Indian National Cyber Registry, India Today, 5 November 2015, last retrieved on 2 August 2018 at http://indiatoday.intoday.in/story/ntro-proposes-creation-of-indian-national-cyber-registry/1/517355.html

Any acquisition of cyber tools (both paid and open source) should be accompanied by registration (with a valid ID) to be able to track the use and attribute it to the owner. To achieve this, a global or at least state-controlled national registry should be established and enforced.

1.2.2 Cyber Arms Industry Infrastructure

The cyber markets offer services and products and whatever their cost is, they are always profitable for malicious actors. The latest report from McAfee's Center for Strategic and International Studies[107] estimates that the cost of cybercrime and cyber-espionage is somewhere around 160 billion USD per year.

Depending on the methodology and available resources, the financial estimates of markets and profits vary. According to Markets and Markets, the cybersecurity market is expected to grow from 137.85 billion USD in 2017 to 231.94 billion USD by 2022, at a Compound Annual Growth Rate (CAGR) of 11.0 percent. The major forces driving the illegal cybersecurity market are strict data protection directives and cyber terrorism. Data protection allows the malicious actors to abuse their privacy rights and hide using conventional protocols and tools. The cybersecurity market is growing rapidly because of the growing security needs of the Internet of Things and Bring Your Own Device trends, and the increased deployment of web and cloud-based business applications.[108]

Goods and Services

Among the most highly demanded goods on the market are software vulnerabilities and exploits. When the majority of the population uses the same software, one specific vulnerability can be used against millions of computers, which is very attractive to malicious actors.

Zero-day vulnerabilities and exploits are virtual products that can be easily sold without intermediaries over the internet. The available technologies and methods provide anonymity to those activities at very low costs, as well as technical support services and tutorials. Among the facilitating tools there are offshore VPN servers, custom malware generation and encryption for additional privacy and antimalware evasion, credentials to corporate networks, fake IDs and credit cards,

---

[107] Net losses: Estimating the global cost of cybercrime, Center for Strategic and International Studies, June 2014, last retrieved on 14 July 2018 at https://www.sbs.ox.ac.uk/cybersecurity-capacity/system/files/McAfee%20and%20CSIS%20-%20Econ%20Cybercrime.pdf

[108] Cybersecurity Market by Solution, Service, Security Type, Deployment Mode, Organization Size, Vertical, and Region - Global Forecast to 2022, last retrieved on 14 October 2018 at https://www.marketsandmarkets.com/PressReleases/cyber-security.asp

custom-made and preconfigured hacking devices (USB drives, rogue routers, battery powered microcomputers).

The costs of the services vary depending on the quality. The 2018 McGuire' report on International Cyber Crime[109] found numerous examples of services and products for sale on various on-line platforms:

- Zero-day Adobe exploits, up to 30,000 USD
- Zero-day iOS exploit, 250,000 USD
- Malware exploit kit, 200-600 USD per exploit
- Blackhole exploit kit, 700 USD for a month's leasing, or 1,500 USD for a year
- Custom spyware, 200 USD
- SMS spoofing service, 20 USD per month.

In reality these prices may be higher or lower depending on the needs, urgency and market offers.

Markets

The cyber arms markets can be white, grey and black based on the level of legality they operate with.

White markets sell official goods and services and, as an example, offer rewards for discovered vulnerabilities and submit them to the original developers, develop commercial vulnerability scanners, provide consulting services regarding cyber operations and defence. Developers may reward security researchers for reporting the vulnerabilities under the so called "bug bounty" program in a limited scope and under specific rules of engagement. On average, the prices reported until 2018[110] were less than 3,000 USD but special offers up to 100,000 USD (and sometimes up to 2 mil USD[111]) were made for certain vulnerabilities based on the type, criticality and nature of the affected software. Some companies prefer contract penetration testing experts to discover internal and external vulnerabilities for their personal use, without disclosing them further through the public domain.

Grey markets are unofficial markets of goods that have not been obtained from official suppliers. They usually require the use of the third parties to hide the traces of their transactions, and their buyers normally include clients from the private sector, brokers who resell vulnerabilities,

---

[109] Hyper-connected web of profit emerges, as global cybercriminal revenues hit $1.5 trillion annually, Bromium Report, 20 April 2018, last retrieved on 16 October 2018 at https://www.bromium.com/press-release/hyper-connected-web-of-profit-emerges-as-global-cybercriminal-revenues-hit-1-5-trillion-annually/

[110] HackerOne, https://www.hackerone.com/; BugCrowd, https://www.bugcrowd.com/

[111] Program Overview, Zerodium, last retrieved on 13 January 2019 at https://www.zerodium.com/program.html

and investigation agencies. Information about these markets is confidential and available through requests to special agents.

Black markets provide a wider variety of services as they are not constrained by legal rules and regulation. The prices vary but tend to be much higher than at the white market.

At the grey and black cyber markets most purchases are made with cryptocurrencies, which are among the largest unregulated financial markets in the world with around 72 billion USD of illegal activities per year. Cryptocurrencies are transforming the way black markets operate by enabling the "black e-commerce". It is claimed, that approximately one-quarter of the bitcoin users and one-half of bitcoin transactions are associated with the illegal activity. The illegal share of bitcoin activity declines with the mainstream interest in bitcoin and with the emergence of more opaque cryptocurrencies.[112]

Non-stop conflicts accompany the cyber markets – for profits, reputation, channels of distribution, priority acquisitions, etc. Their supply chain is complex and involves multiple actors organized by hierarchies, including administrators, technical experts, intermediaries, brokers, vendors, witting mules, etc. Their duties and responsibilities vary and while zero-day exploits can be "found" or developed by subject matter experts only, other exploits can be easily commercialized by almost any person willing to enter the black market. Human brokers and bug bounty platforms do not always require information about the suppliers of the products (vulnerabilities, exploits, etc.), thus these suppliers can remain anonymous.

In 2015, the company Zerodium[113] initiated the acquisition of "high-risk vulnerabilities" and announced a new bounty program. The company published the formats required for vulnerability submissions, and the criteria to determine prices that is, the popularity and complexity of the affected software, and the quality of the submitted exploit – and the prices themselves. This initiative represents a mixture of the transparency offered by the traditional vulnerability reward program and the high rewards offered in the grey and black markets. Software developer companies perceived this new approach as a threat, primarily since very high bounties could cause developer and tester employees to leave their day jobs in exchange for freelance bug bounty hunting on a legal basis.

As a platform for the grey and black markets, there exist several implementations:

---

[112] S. Foley, J.R. Karlsen, T.J. Putnis, Sex, Drugs, and Bitcoin: How Much Illegal Activity Is Financed Through Cryptocurrencies? 14 December 2018, last retrieved on 18 December 2018 at https://papers.ssrn.com/sol3/papers.cfm?abstract_id=3102645

[113] Program Overview, Zerodium, last retrieved on 5 January 2019 at https://zerodium.com/program.html

- Physical in-person trade (no computer networks involved)
- Closed access webservers (not indexed by search engines and inaccessible without specific knowledge of the internet address and user/password combination)
- Darknet markets (encrypted overlay over the traditional Internet)

Darknet Markets

Darknet markets are unregulated platforms, an encrypted overlay over the traditional Internet, using specific protocols and software (Tor, i2p, freenet, etc). These markets also use encryption and privacy mechanisms such as "off-the-record" messaging protocol, stateless operating systems, anonymous VPNs, and cryptocurrencies. Though providing a high level of anonymity, most of them are easily accessible without any advanced technical knowledge.

The darknet markets are numerous, such as Silk Road, AlphaBay, Hansa (taken over through law enforcement operations in 2013 and 2017), interconnected and growing from one another. They advertise on established forums such as Dream Market, TradeRoute, House of Lions and Wall Street Market, etc. These forums advertise their products and services, and the dark web sites facilitate the payment. Some of the darknet carding and Automated Vending Cart sites help in commit fraud with the payment cards, and even providing "how to" tutorials.

The Cyber Arms Bazaar, a darknet market operating out of various Eastern European countries, trafficking crimeware and hacking tools, has been in action since at least the year 2000. Tom Kellermann, Chief Cybersecurity Officer of Trend Micro, estimates over 80 percent of financial sector cyberattacks could be traced back to the bazaar, with retail cyberattacks not far behind.[114]

Darkode (which was taken down in 2015), a cybercrime forum and black marketplace described by Europol as the most prolific English-speaking cybercriminal forum to date, was a one-stop, high-volume shopping venue for some of the world's most prolific cyber criminals. This underground, password-protected, online forum launched in 2007, was a meeting place for those interested in buying, selling, and trading malware, botnets, stolen personally identifiable information, credit card information, hacked server credentials, and other data and software that facilitated complex cybercrimes all over the globe.[115] In July 2015, the

---

[114] C. Bennett, "Feds search for ways to impede 'cyber bazaar'", 15 March 2015, last retrieved on 16 October 2018, http://thehill.com/policy/cybersecurity/235726-feds-search-for-ways-to-impede-cyber-bazaar

[115] *Cyber Criminal Forum Taken Down,* Members Arrested in 20 Countries, *15 July 2015, last retrieved on 16 August 2018 at* https://www.fbi.gov/news/stories/cyber-criminal-forum-taken-down/cyber-criminal-forum-taken-down

cybercrime forum Darkode was seized by FBI and various members were arrested in "Operation Shrouded Horizon".[116] (See Figure 1.3)

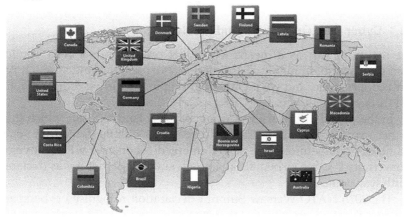

Figure 1.3: The Shrouded Horizon investigation against the Darkode cyber criminal forum involved law enforcement agencies in 20 countries.[117]

The Real Deal, a darknet website and part of the cyber-arms industry, was reported to have been selling all sorts of cybercrime products and services, including code and zero-day software exploits, remote access trojans, cracked accounts, datasets, etc.[118] It was taken down in November 2016.

The above review of the cyber arms industry shows that the industry is prospering, adjusting to and being reinforced by new emerging technologies, and proving an increasing range of products and services to developers, distributers and end users.

---

[116] A. Buncombe Darkode: FBI shuts down notorious online forum and cracks 'cyber hornet's nest of criminal hackers', 15 July 2015, last retrieved on 16 August 2018 at https://www.independent.co.uk/news/world/americas/darkode-fbi-shuts-down-notorious-online-forum-and-cracks-cyber-hornets-nest-of-criminal-hackers-10391734.html

[117] Ibid

[118] J. Cox, The Mysterious Disappearance, and Reappearance, of a Dark Web Hacker Market, 29 July 2015, last retrieved on 16 August 2018 at https://motherboard.vice.com/en_us/article/8qxzap/the-mysterious-disappearance-and-reappearance-of-a-dark-web-hacker-market

1.3 Cyberspace as a Military domain

> *"One hundred victories in one hundred battles*
> *is not the most skilful.*
> *Seizing the enemy without fighting is the most skilful."*
>
> Sun Tzu, "The Art of War", 6th c. BC

1.3.1 Cyber War, Operations and Incidents

The nature of war as a means of conflict resolution has been applied from the dawn of humanity, fighting for lands, seas, air and space. It has evolved with time, and we are now fighting in the virtual dimension, for and through cyberspace.

In the past the primary war domains were land and sea, which then further expanded to air, and then to space - a unique domain of space technology competition. The newly emerged cyberspace with its overall connectivity became so beneficial and advantageous, that it was declared a fifth war domain.

The 2016 NATO Warsaw Summit declaration recognised cyberspace as a domain of operations, and that "cyber-attacks could be as harmful as a conventional attack and present a clear challenge to the security of the Alliance".[119] Similarly, The Law of War Manual recognises the cyberspace as an operational domain, just like land, sea, air and space domains.[120] The Joint Publication 3-12 "Cyberspace Operations" as of 8 June 2018 defines "cyberspace" as "A global domain within the information environment consisting of the interdependent networks of information technology infrastructures and resident data, including the Internet, telecommunications networks, computer systems, and embedded processors and controllers."[121]

In the broad sense of the word, cyberspace is defined by Joseph Nye as the "Internet of networked computers but also intranets, cellular technologies, fibre optic cables, and space-based communications"[122]. Cyberspace refers to not only "all of the computer networks in the world"

---

[119] Warsaw Summit Communiqué Issued by the Heads of State and Government participating in the meeting of the North Atlantic Council in Warsaw 8-9 July 2016, last retrieved on 15 July 2018 at https://www.nato.int/cps/en/natohq/official_texts_133169. htm

[120] Law of War Manual, 2015, updated version, 2016, last retrieved on 15 July 2018 at https://www.defense.gov/Portals/1/Documents/pubs/DoD%20Law%20of%20 War%20Manual%20-%20June%202015%20Updated%20Dec%202016.pdf?ver=2016 -12-13-172036-190

[121] Joint Publication 3-12, Cyberspace Operations, GL-4, 8 June 2018, last retrieved on 7December 2018 at https://fas.org/irp/doddir/dod/jp3_12.pdf

[122] J. Nye, The Future of Power, 2011, Public Affairs, New York, p. 19

but also to "everything they connect and control"[123], hence the potential risk to a nation's infrastructure given the fact that these systems are mostly dependent on Internet networks. And "cyber warfare is best understood as a new but not entirely separate component of this multifaceted conflict environment".[124]

Unlike the term "war" which has been clearly defined for conventional warfare activities, the term "cyber war" is still under discussion and several definitions have been proposed, with no single definition being widely adopted internationally. The online Oxford dictionary defines it as the use of computer technology to disrupt the activities of a state or organization, especially the deliberate attacking of information systems for strategic or military purposes.[125]

Opinions also differ from the negation of "feasibility of waging war by cyber means altogether"[126,127] to its full recognition under the justification that "modern societies" (and the military) have become so dependent on ICT-infrastructures that a disruptive attack on key systems could lead to war-like consequences or decisive advantage in warfare engagements.[128, 129,130,131]

---

[123] Richard A. Clarke and Robert K. Knake, 2010. Cyber War: The Next Threat to National Security and What to Do About It, Harper Collins, New York, p. 70

[124] On Cyber Warfare, P. Cornish et al., Chatham House, 2010, last retrieved on 9 December 2018 at https://www.chathamhouse.org/publications/papers/view/109508

[125] Online Oxford dictionary, last retrieved on 15 October 2018 at https://en.oxforddictionaries.com/definition/cyberwarfare

[126] E. Gartzke, 2013. The Myth of Cyberwar: Bringing War in Cyberspace Back Down to Earth, International Security, 38 (2): 41-73. doi: 10.1162/SEC_a_00136 (Cited from S. Steiger et al., Conceptualising conflict in cyberspace, Journal of Cyber Policy, Chatham House, 2018, p. 77)

[127] T. Rid, 2013. Cyber War Will Not Take Place. Oxford University Press, Oxford, UK. (Cited from S. Steiger et al., Conceptualising conflict in cyberspace, Journal of Cyber Policy, Chatham House, 2018, p. 77)

[128] R.A. Clarke and R.K. Knake, 2010. Cyber War: The Next Threat to National Security and What To Do About It. Ecco, New York. (Cited from S. Steiger et al., Conceptualising conflict in cyberspace, Journal of Cyber Policy, Chatham House, 2018, p. 77).

[129] J. Stone, 2013. Cyber War Will Take Place! Journal of Strategic Studies 36 (1): 101-108. doi: 10.1080/01402390.2012.730485. (Cited from S. Steiger et al., Conceptualising conflict in cyberspace, Journal of Cyber Policy, Chatham House, 2018, p. 77)

[130] R. Stiennon, 2015. There will be Cyberwar: How to move to Network-centric Warfighting Set the Stage for Cyberwar. Birmingham, M: IT-Harvest Press. (Cited from S. Steiger et al., Conceptualising conflict in cyberspace, Journal of Cyber Policy, Chatham House, 2018, p. 77)

[131] E. Goldman and M. Warner, 2017. Why a Digital Pearl Harbor Makes sense... and is Possible. *In*: Understanding Cyber Conflict, 14 Analogies, edited by G. Perkovich and A. Levite, 147-157. Georgetown University Press, Washington D.C.. (Cited from S. Steiger et al., Conceptualising conflict in cyberspace, Journal of Cyber Policy, Chatham House, 2018, p. 77)

The general understanding is that cyberwarfare[132] is the use or targeting in a battle space or warfare context of computers, online control systems and networks. It involves both offensive and defensive operations pertaining to the threat of cyberattacks, espionage and sabotage.

Cyber warfare can enable actors to achieve their political and strategic goals without the need for armed conflict.[133] Though there has not been any official declaration of a cyber war, many governments have voiced the disruptive consequences of on-going hostile cyber activities, including criminal cyberattacks. These two national conflicts, however, could be considered as the most relevant to cyber warfare – the 2007 cyberattacks against Estonia and the 2008 combined cyber and kinetic attacks against Georgia due to the "real-life" implications of having communications disrupted during kinetic combat.[134] Both of them involved nation states and called for external support.

In 2010, Spokesperson Mike McConnell, former Director of National Intelligence, and a Senior Executive for a defence contractor, wrote for Washington Post: "The United States is fighting a cyber-war today, and we are losing. It's that simple."[135] And in 2011, the United States[136] warned that "when warranted, it will respond to hostile acts in cyberspace as it would do to any other threat to the country" thus recognising the potential of an overall high level of cyber threats to the state equal to a war.

In the absence of a commonly accepted definition, there is a growing tendency to avoid the strong term "cyber war", replacing it with more neutral and politically acceptable "cyber conflicts", "cyber operations", "cyber incidents". In 2016, Craig and Valerian (Cardiff University) in "Conceptualising cyber arms races"[137] wrote: cyber conflict can be defined as "the use of computational technologies in cyberspace for malevolent

---

132 L.H. Newman, Menacing malware shows the dangers of industrial system sabotage, 18 January 2018, last retrieved on 16 October 2018 at https://www.wired.com/story/triton-malware-dangers-industrial-system-sabotage/?CNDID=50121752

133 On Cyber Warfare, P. Cornish et al., Chatham House, 2010, last retrieved on 9 December 2018 at https://www.chathamhouse.org/publications/papers/view/109508

134 Cybersecurity and Cyberpower: Concepts, Conditions and Capabilities for Cooperation for Action within the EU, SEDE, 2011, last retrieved on 9 December 2018 at http://www.oiip.ac.at/fileadmin/Unterlagen/Dateien/Publikationen/EP_Study_FINAL.pdf

135 Mike McConnell on how to win the cyber-war we're losing, 25 February 2010, last retrieved on 15 July 2018 at http://www.washingtonpost.com/wp-dyn/content/article/2010/02/25/AR2010022502493.html?noredirect=on

136 Barack Obama, International Strategy for Cyberspace: Prosperity, Security, and Openness in a networked World, 14 May 2011, last retrieved on 9 June 2018 at https://obamawhitehouse.archives.gov/sites/default/files/rss_viewer/international_strategy_for_cyberspace.pdf

137 A. Craig and Dr B. Valerian, Conceptualising cyber arms races, 2016, last retrieved on 9 June 2018 at http://www.ccdcoe.org/cycon/2016/proceedings/10_craig_valeriano.pdf

and/or destructive purposes in order to impact, change, or modify diplomatic and military interactions between entities".[138]

Cyber conflicts vary as per their intensity and there exist datasets for cyberattacks with their qualitative major characteristics. For examples, since 2006 the Center for Strategic and International Studies has kept a list of events, the Cyber Operations Tracker (2017) records state-sponsored attacks.

However, Steiger et al.[139] argue that "until now the concept of cyber conflict has been mostly reduced to unilateral cyberattacks, such as the DDoS attacks on Estonia in 2007 or the Olympic Games operation against the Iranian nuclear programme, or to instances in which conventional warfare has involved cyber tools". They define cyber conflict as "incompatibility of stated intentions between actors which guide their use of computer technologies to harm the other"[140] and, based on the studies of conventional conflicts, propose ways to develop criteria to reflect the intensity of cyberattacks through a structured scoring of potential on and offline effects. They highlight the need for a more comprehensive approach to global data base of cyberattacks, and their classification, in order to better measure the intensity and processes of conventional conflicts and their resonance in cyber space and vice versa. To probe the potential for cyber wars and to address the rising number of attacks on critical infrastructure, they propose to measure the physical effects, differentiating between "temporal" and "spatial" dimensions.

The use of technology to both control and disrupt the flow of information has been generally referred to by several names: information warfare, electronic warfare, cyberwar, netwar, and Information Operations (IO). Currently, IO activities are grouped by the Department of Defense (DOD) into five core capabilities: (1) Psychological Operations, (2) Military Deception, (3) Operational Security, (4) Computer Network Operations, and (5) Electronic Warfare.

Current U.S. military doctrine for IO now places increased emphasis on Psychological Operations, Computer Network Operations, and Electronic Warfare, which includes use of non-kinetic electromagnetic pulse (EMP) weapons, and nonlethal weapons for crowd control. However, as high technology is increasingly incorporated into military functions, the boundaries between all five IO core capabilities are becoming blurred.

The DOD has noted that military functions involving the electromagnetic spectrum take place in what is now called the cyber domain, similar to air, land, and sea. This cyber domain is the responsibility of the new Air Force Cyber Command and includes cyberwarfare, electronic warfare, and protection of U.S. critical infrastructure networks that support telecommunications systems, utilities, and transportation".

---

[138] B. Valeriano and R.C. Maness, Cyber War versus Cyber Realities, New York: Oxford University Press, 2015, p. 32

[139] S. Steiger et al., Conceptualising conflict in cyberspace, Journal of Cyber Policy, Chatham House, 2018, p. 78

[140] Ibid

Source: Wilson, Clay. 2007. Information operations, electronic warfare, and cyberwar capabilities and related policy issues. [Washington, D.C.]: Congressional Research Service, Library of Congress. [141]

The Law of War Manual[142], released by the US Department of Defence in 2015, and updated in May 2016, addresses "how law of war principles and rules apply to relatively novel cyber capabilities and the cyber domain", though mentioning that this area is not well settled and is likely to continue to develop. Without mentioning the term "cyber war", it reduces its definition to cyber operations: "cyberspace operations may be understood to be those operations that involve "the employment of cyberspace capabilities where the primary purpose is to achieve objectives in or through cyberspace." [143]

It defines "cyber operations" [144] as activities that:

- use cyber capabilities, such as computers, software tools, or networks;
- have a primary purpose of achieving objectives or effects in or through cyberspace.

The technical success of a cyber operation is determined by the qualification of the operators, available software, computational power of hardware, and environmental factors.

Cyber operations target computer systems to disrupt, deny, degrade, or destroy information in computers and computer networks, or the computer and networks themselves. They can be also used to advance force operations. They may include reconnaissance, seizure of opportunity positions, pre-deployment of capabilities of weapons (e.g., mapping a network), seizure of supporting positions (e.g., securing access to key network systems or nodes), and pre-emplacement of capabilities or weapons (e.g., implanting cyber access tools or malicious code). In addition, cyber operations may be a method of acquiring foreign intelligence unrelated to specific military objectives, such as understanding technological developments or gaining information about an adversary's military capabilities and intent. [145]

---

[141] C. Wilson, Information Operations, Electronic Warfare, and Cyberwar: Capabilities and Related Policy Issues, Report for Congress, 20 March 2007, last retrieved on 10 June 2018 from https://www.history.navy.mil/research/library/online-reading-room/title-list-alphabetically/i/information-operations-electronic-warfare-and-cyberwar.html

[142] Law of War Manual (Section 16.1.2), updated on May 2016, last retrieved on 9 June 2018 from https://www.defense.gov/Portals/1/Documents/pubs/DoD%20 Law%20of%20War%20Manual%20-%20June%202015%20Updated%20Dec%202016. pdf?ver=2016-12-13-172036-190

[143] Joint Publication 3-0, Joint Operations, 11 August 2011

[144] Law of War Manual (Section 16.1.2) Retrieved on 18 July 2018 from https://www. defense.gov/Portals/1/Documents/pubs/DoD%20Law%20of%20War%20Manual%20 -%20June%202015%20Updated%20Dec%202016.pdf?ver=2016-12-13-172036-190

[145] Ibid

Considering that the term "war" has evolved by itself, and now includes a war against such global issues as drugs, human trafficking, and terrorism, we can assume that offensive cyber activities can be termed as a cyber war. It has its clear advantages, such as a covert and stealthy character, remote operation and absence of psychological barriers for combatants, safety for human operators in military missions, access to hard-to-reach facilities and systems, etc. The US Air Force has had Information Warfare Squadrons since the 1980s with the official mission "To fly, fight and win... in air, space and cyberspace". [146]

> **US 'launched cyber-attack on Iran weapons systems' [147]**
> They were using "destructive 'wiper' attacks", [...] using tactics such as "spear phishing, password spraying and credential stuffing" in a bid to take control of entire networks. Iran has also been trying to hack US naval ship systems, the Washington Post reported. The attack had been planned for several weeks, the sources told US media outlets, and was suggested as a way of responding to the mine attacks on tankers in the Gulf of Oman.
> It was aimed at weapons systems used by Iran's Islamic Revolutionary Guard Corps (IRGC), which shot down the US drone last Thursday and which the US says also attacked the tankers.
> Both the Washington Post and AP news agency said the cyber-attack had disabled the systems. The New York Times said it was intended to take the systems offline for a period of time.

In this book, we recognize the term "cyber war" as a broad spectrum of short and long term strategically planned cyber offensive operations, with a compelling effect and high impact offensive goals. We also recognize that there might be numerous classifications of types of cyber wars, but in general, a cyber war may be either a cyber war *per se* (intelligence gathering and/or sabotage), or a cyber war as a part of a hybrid war, engaging all military domains.

Cyber war *per se* has the following characteristics:

- Covert and speedy character of cyber operations
- Absence of clear protocols
- Increased risk and uncertainty, both for attackers and defenders
- Difficulties in the attribution

> **DIME Model [148]**
> Cyber war can influence all instruments of national power: diplomatic, information, military, and economic.

---

[146] U.S. Air Force, https://www.airforce.com/mission

[147] US 'launched cyber-attack on Iran weapons systems', 23 June 2019, BBC, https://www.bbc.co.uk/news/world-us-canada-48735097

[148] R. Hillson, The DIME/PMESII Model Suite Requirements Project, 2009, last retrieved on 15 October 2018, https://www.nrl.navy.mil/content_images/09_Simulation_Hillson.pdf

Diplomatic influence is based on the actions between states based on official communications. It can go through government organizations, international treaty organizations such as the North American Treaty Organization (NATO), economic groups like the Group of Twenty Finance Ministers and Central Bank Governors (G20), or law enforcement agencies.

The political environment can be influenced through **Information**, strategic communication, news and popular media, international opinion, social media sites, and Open Source Intelligence; the information cyber war can gain power over the minds of the key national actors.

Military engagement is the final political option, but today we see the full spectrum of its engagement, from unconventional warfare, peacekeeping, humanitarian assistance, nation-building, and finally large-scale combat operations.

Economic power comes from the influence of trade, incentives such as embargos and free trade zones and direct support for examples, aid packages or sale of surplus DoD equipment. All these factors can be applied to effect behaviours in cyber warfare.

Cyber war as a part of a hybrid war is a more complicated phenomenon. The hybrid warfare, using mixed methods, such as propaganda, deception, sabotage and other non-military tactics have long been used to destabilise adversaries. What is new about the attacks seen in recent years is their speed, scale and intensity, facilitated by rapid technological change and global interconnectivity.[149]

All technological achievements are now employed in warfare operations, and this will only escalate in the future. Defence strategies will be built with multi war domain consideration and defence technics will require a joint expertise of professionals in all these areas. The Brussels Summit Declaration, issued by the Heads of State and Government participating in the meeting of the North Atlantic Council in Brussels 11-12 July 2018, warns: "We face a dangerous, unpredictable, and fluid security environment, with enduring challenges and threats from all strategic directions; from state and non-state actors; from military forces; and from terrorist, cyber, and hybrid attacks. [...] Hybrid activities ... aim to create ambiguity and blur the lines between peace, crisis, and conflict."[150]

Hybrid threats are diverse and ever-changing, and the tools used range from fake social media profiles to sophisticated cyberattacks, and all the way to the overt use of military force and everything in between. Hybrid influencing tools can be employed individually or in combination, depending on the nature of the target and the desired outcome. As a

---

[149] NATO response to cyber threats, 17 July 2018, last retrieved on 20 July 2018 at https://www.nato.int/cps/en/natohq/topics_156338.htm

[150] Brussels Summit Declaration, Issued by the Heads of State and Government participating in the meeting of the North Atlantic Council in Brussels 11-12 July 2018, last retrieved on 8 December 2018 at https://www.nato.int/cps/en/natohq/official_texts_156624.htm#21

necessary consequence, countering hybrid threats must be an equally dynamic and adaptive activity, striving to keep abreast of variations of hybrid influencing and to predict where the emphasis will be next and which new tools may be employed.[151]

Brussels Summit Declaration

Article 21: "Our nations have come under increasing challenge from both state and non-state actors who use hybrid activities that aim to create ambiguity and blur the lines between peace, crisis, and conflict. While the primary responsibility for responding to hybrid threats rests with the targeted nation, NATO is ready, upon Council decision, to assist an Ally at any stage of a hybrid campaign. In cases of hybrid warfare, the Council could decide to invoke Article 5 of the Washington Treaty, as in the case of armed attack. We are enhancing our resilience, improving our situational awareness, and strengthening our deterrence and defence posture. We are also expanding the tools at our disposal to address hostile hybrid activities. We announce the establishment of Counter Hybrid Support Teams, which provide tailored, targeted assistance to Allies, upon their request, in preparing for and responding to hybrid activities."[152]

Brussels Summit Declaration, Issued by the Heads of State and Government participating in the meeting of the North Atlantic Council in Brussels 11-12 July 2018

The evolution of weapons, strategies and war actors per se, has led to the transition of war to the next, so called fourth generation warfare (4GW).[153,154] The absence of a proper definition reflects its character. With its multiple transitional stages, it has no clear distinction between war and peace, no clear rules and procedures, no defined battlefields, no clear separation of civil or military participants with headquarters and command and control systems, and no concrete timelines or concluded peace treaties. The self-learning weapons, artificial intelligence and autonomous robotic systems are penetrating the conventional battlefields, modernizing tactics, and disguising the military's activities making them more covert and unpredictable.

Cyber operations will be an essential part of the 4GW with full spectrum of their covert capacities and considerably enhanced by other technological advancements. One of the emerging threats is the combination of cyber and electromagnetic attacks. The equipment for cyber operations can be

---

[151] Cooperating to counter hybrid threats, NATO Review, 23 November 2018, last retrieved on 8 December 2018 https://www.nato.int/docu/review/2018/Also-in-2018/cooperating-to-counter-hybrid-threats/EN/index.htm

[152] Ibid

[153] William S. Lind, COL Keith Nightengale, CPT John F. Schmitt, COL Joseph W. Sutton, and LTC Gary I. Wilson, "The Changing Face of War: Into the Fourth Generation, Marine Corps Gazette (October 1989): 22-26. Last retrieved on 8 December 2018 at https://www.mca-marines.org/gazette/2001/11/changing-face-war-fourth-generation

[154] William S. Lind, Understanding Fourth Generation War, Military Review, 2004, last retrieved on 8 December 2018 at http://www.au.af.mil/au/awc/awcgate/milreview/lind.pdf

affected by electromagnetic pulse (EMP) weapons, the transmissions can be jammed or intercepted.

Electromagnetic Attacks

- Electromagnetic pulse (EMP) and cyberattack..., operating in tandem, can disable not just a significant portion of the electrical grid and critical infrastructure, but also the network-centric military response to such an attack. If a high-altitude EMP attack were paired with both a large-scale cyberattack and a biological attack, the resulting challenge to interagencies could surpass anything the interagencies are currently structured or equipped to respond to.
- Current preparedness and response plans focus primarily on one weapons of mass destruction (WMD) attack mode at a time. However, an EMP and cyberattack would amplify the effects of a biological attack and vice-versa. The ramifications of such a combination of attacks are staggering.
- Detection of biological agents could be disabled after an EMP and cyberattack because electronic healthcare-surveillance systems would be no longer operational and could no longer process and exchange information among agencies. Laboratories would no longer receive or process suspected specimens to identify potentially hazardous biological agents. Without a timely response, the spread of disease in a population may not be contained during its early stages and could lead to outbreaks and epidemics. Without the ability to detect biological agents, public health officials cannot initiate timely treatment and preventive measures.[155]

Patricia Rohrbeck, Concurrent Biological, Electromagnetic Pulse and Cyber-attacks: The Ultimate Interagency Response Challenge, 2017

As per the WEF 2018 Global Risks Perception Survey[156], the likelihood of cyberattacks occupies the third place after non-man disasters like "Extreme weather events and natural disasters", and their impact – the sixth place.

1.3.2 Actors in Cyber Conflicts

Behind any cyber operation, there is a human factor. In the cyber war, these can be both state and non-state actors, such as terrorist groups, companies, political or ideological extremist groups, hacktivists, transnational criminal organizations, or even non-human autonomous actors. As mentioned previously, the modern cyber war conditions allow any of them to be active participants of cyber offensive activities.

---

[155] P. Rohrbeck, Concurrent Biological, Electromagnetic Pulse and Cyber-attacks: The Ultimate Interagency Response Challenge, 2017, last retrieved on 8 December 2018 at http://thesimonscenter.org/wp-content/uploads/2017/05/IAJ-8-2-2017-pg53-61.pdf

[156] WEF 2018 Global Risks Report, last retrieved on 16 July 2018 at http://www3.weforum.org/docs/WEF_GRR18_Report.pdf

---

Actors in Cyber Conflicts and their Goals

State actors:
- Government: to deter enemies, guard nation sovereignty, attack an enemy country, etc.
- Military: to offend and defend
- Intelligent agencies: to gather information, assist in investigation, etc.
- Political forces: to enforce political changes

Non-state actors:
- Terrorists: to disrupt, damage
- Criminals: to gain financial profits
- Industries and corporations: to gain advantage over competitors, financial profits and markets, ensure leading roles in specific industrial fields
- Hacktivists: to protest against social inequity, support points of view
- Individuals: to protest for personal issues, for fun

Emerging threat (non-humans):
- AI malware: accidental
- Web crawlers: accidental

Executors:
Military cyber security and cyber operations experts are considered to be the highest skilled experts possible in the area.

State actors, with government and military support and resources, have considerable advantage in the cyber operations. However, they are restricted by rules and regulations both at the national and international levels.

Non-state actors, e.g., individuals, hacktivists, criminals, criminal organizations, or any other individuals or groups, who can conduct cyber operations on national or multinational levels, have more limited resources but are more flexible in actions and geographical location. They may be engaged by states and might not necessarily respect the law. It is the task of the state to control the malicious activities of the non-state actors on its territory. Depending on the legal regulations and enforcement inside the state, the level of state control varies.

The emerging threat is represented by a group of **autonomous actors**, i.e. enhanced malware able to function separately from a human operator. When malware is released, or just disconnected from its command and control structure, it may continue to carry out designated functions independent of any outside control. This has been the case since the very first pieces of malware were seen outside of controlled environments, but now with AI enhanced agents the issue is much more critical.

With the advanced technologies, the speed of cyber actions, whether offensive or defensive in nature, is limited only by the speed of the networks and systems in which they take place. This means that engagements in cyber warfare can take place at speeds exceeding human control capabilities. The human factor remains the weakest link, and it is through the human command that the response to a cyberattack has a potential to be escalated disproportionally with the use of conventional weapons.

1.3.3 Cyberattack Scenario

Cyberattacks can be short or long, demonstratively visible or stealthily covert, ordered and funded by state or non-state actors. They have strategic objectives and specific targets, and despite their diversity, follow certain development schemes, presented in Chapter 1 (1.1.2). These schemes can be implemented fully or partially, depending on the goal and accompanying activities. Ideally, the following stages constitute a full-scale cyberattack operation.

Reconnaissance

The reconnaissance phase is the essential step at the beginning of every operation, allowing for proper planning based on collected preliminary information. Information gathering covers all valuable data about the target, its strengths and weaknesses, including physical locations, entry points, number of computers connected to the internet, public IP addresses of the servers, etc.

Scanning

During the scanning phase, the targeted system is examined closely with special attention for potential vulnerabilities. It is worth noting here, that the web applications, especially those developed with tough deadlines and without strong security testing, provide one of the easiest ways to survey a remote system, and potentially open the door into the operating system.

Scanning collects specific information about the operating systems such as versions, uptime, system language, etc. Based on the collected details, the engagement plans are developed with multiple options.

Initial Breach and Exploitation

The initial breach can be performed through a variety of methods, such as password guessing, application exploitation, web application exploitation, planting a rogue network device, etc. (see 1.1.2 Cyber offence tools and techniques).

In the case of a high level operation, the attackers try to identify unknown vulnerabilities to ensure the absence of mitigation measures or patching. The same refers to the used exploits, which preferably should be unavailable to the general public.

Privilege Escalation

Privilege escalation is the process of gaining higher level access rights to the target system, by exploiting the system vulnerabilities using already existing partial access. It can be accomplished through the exploitation of the system kernel or by taking advantage of misconfigurations or insecurely set configurations and local applications that are operating with higher permissions. In misconfigured systems the low-privileged user may have the ability to act as an administrator directly and run specific commands, which might be misused by an attacker. The attackers may be able to access and modify custom made scripts or shell scripts that are not secured properly, in order to pass operating system commands through them or gain direct access to an operating system shell.

Persistence

After gaining access to the system, the attacker's goal is to establish persistence. They can create additional user accounts, open services on additional ports, install command and control software (remote administration tools), or place a backdoor in a regularly active application.

In addition, the attackers may patch or fix the vulnerabilities through which they were able to gain access. This will ensure protection against other malicious groups and competitors and create confusion in case of potential investigation.

Exfiltration

After receiving sufficient access to the target network, the next step is to locate the requested data and extract it to a neutral environment accessible from another location. The wide variety of tools exist for data exfiltration, e.g., built-in tools for data transfer, the purpose-built tools and protocols to data transmission, cloud services, etc.

Assault

When the goal is to cause damage or destroy a system or network, the attackers can interrupt the processes, stop controls, alter targets, over-utilize resources, move or manipulate the files, trigger alarms, etc. This might cause disruption of the planned activities, panic in personnel, change of transport route and produce other unexpected effects.

Spoofable Sirens [157]
At exactly noon on the first Tuesday after Balint Seeber moved from Silicon Valley to San Francisco in late 2015, the Australian radio hacker and security researcher was surprised to discover a phenomenon already known to practically every other resident of the city: a brief, piercing wail that rose and then fell, followed by a man's voice: "This is a test. This is a test of the outdoor warning system. This is only a test."

---

[157] This radio hacker could hijack citywide emergency sirens to play any sound, 10 April 2018, last retrieved on 8 October 2018 at https://www.wired.com/story/this-radio-hacker-could-hijack-emergency-sirens-to-play-any-sound/

The next week, at exactly the same time, Seeber heard it again. A few weeks after that, Seeber found himself staring up from his bicycle at a utility pole in the city's SoMa neighbourhood, examining one of the more than 100 sirens that produced that inescapable emergency test message around the city. At the top, he noticed a vertical antenna; it seemed to be receiving signals via radio, not wires. The thought came to him: Could a hacker like him hijack that command system to trigger all the sirens around the whole city at will, or to use them to broadcast even more alarming sounds?...

After two-and-a-half years of patiently recording and reverse-engineering those weekly radio communications, Seeber has indeed found that he or anyone with a laptop and a 35 USD radio could not only trigger those sirens, as unknown hackers did in Dallas last year. They could also make them play any audio they choose: false warnings of incoming tsunamis or missile strikes, dangerous or mass-panic-inducing instructions, 3 am serenades of death metal or Tony Bennett. And he has found the same hackable siren systems not only in San Francisco but in two other cities, and hints that they may be installed in many more.

Covering Tracks

This stage is one of the most important in the attack, as it ensures covertness and deletes any digital forensic evidence. For example, the attackers change timestamps to reflect the original time before any files were modified, clean up any offensive tools, remove or alter log entries, patch the vulnerabilities they exploited, close open ports, reconfigure firewall rules, etc. On the contrary, false tracks can be intentionally left to direct investigation to some desired track.

During all these stages, even before the first reconnaissance, obfuscation measures are employed. To prevent tracking of the attackers' location, various proxies or intervening machines can be used as an intermediary connection before attacking; IP spoofing, or other methods are used to disguise the unique features of the attack platform. The obfuscation tools do not guarantee full anonymity, but provide a sufficient level of stealth, untraceable with modern forensic techniques.

1.3.4 Deterrence, Coercion and Compellence

In the discussion of the specific characteristics of a cyber war, the question of its efficiency as compared to the conventional war remains open. Cyber tools alone are unable to lead to the overall concession of the adversary. Should this be the final goal, they must be integrated with the other military weaponry and extended coercive measures, such as military threats, social and economic sanctions, propaganda, diplomatic campaigns, etc.

Cyberspace has its own restraints and limits, which consequently limit the coercive potential of cyber weapons. Gartzke and Lindsay[158]

---

[158] E. Gartzke and J. Lindsay, Thermonuclear Cyberwar, Research paper, Oxford University Press, Journal of Cybersecurity, 3(1), 2017, 37–48

highlight that "the potential of cyberspace is more limited than generally appreciated, but it is not negligible, especially when exploited in conjunction with other forms of power such as military force."

However, we consider it essential to review how concepts of deterrence, coercion and compellence could be applied in a cyber war, and across domains.

> Coercion is the action or practice of persuading someone to do something by using force or threats.
>
> Oxford dictionary [159]

Hodgson [160] points out that any discussion of coercion begins with Thomas Schelling's classic writing on the topic, particularly his work "Arms and Influence" [161]. Schelling described two forms of coercion: active coercion, or compellence, and passive coercion, or deterrence. The former involves the active use of force in some form to compel action by another, while the latter involves the threatened use of force to motivate an action or restraint from an action.

Cyber capabilities give its user, "the ability to control and apply typical forms of control and domination of cyberspace". [162] Cyber coercion is used as a means to exert influence or pressure on others to shape behaviour, deter adverse actions and even compel another actor (either another state, a multinational organization, or even a single individual). [163] The compellent coercion measures may apply psychological influence (uncertainty, rumours about cyber potential of attackers), punishment, denial strategies, etc.

Cyber coercion is harmful before cyber operations, as it can reveal the threat to the opponent. Revelation of a cyber threat in advance that is specific enough to convince a target of the validity of the threat also provides

---

[159] Coercion – Definition, Oxford Dictionaries, last retrieved on 7 September 2018 at https://en.oxforddictionaries.com/definition/coercion

[160] Quentin E. Hodgson, Understanding and Countering Cyber Coercion, RAND Corporation, 2018 10th International Conference on Cyber Conflict 2018, last retrieved on 10 December 2018 at https://ccdcoe.org/sites/default/files/multimedia/pdf/Art%20 04%20Understanding%20and%20Countering%20Cyber%20Coercion.pdf

[161] Schelling Th., Arms and Influence, Yales University, 1966, quoted from Quentin E. Hodgson, Understanding and Countering Cyber Coercion, RAND Corporation, 2018 10th International Conference on Cyber Conflict 2018, last retrieved on 10 December 2018 at https://ccdcoe.org/sites/default/files/multimedia/pdf/Art%2004%20Understanding %20and%20Countering%20Cyber%20Coercion.pdf

[162] B. Valeriano and R. Maness, Cyber War versus Cyber Realities: Cyber Conflict in the International System, Oxford University Press, 2015, p. 28

[163] Quentin E. Hodgson, Understanding and Countering Cyber Coercion, RAND Corporation, 2018 10th International Conference on Cyber Conflict 2018, last retrieved on 10 December 2018 at https://ccdcoe.org/sites/default/files/multimedia/pdf/Art%20 04%20Understanding%20and% 20Countering%20Cyber%20Coercion.pdf

enough information potentially to neutralize it.[164] Cyber operations, especially against nuclear command, control and communication systems (NC3), must be conducted in extreme secrecy as a condition of the efficacy of the attack. Cyber tradecraft relies on stealth, stratagem, and deception.[165] However, a chain of successful cyber operations may result in a required coercive effect, especially with following disruptions and sabotage.

Both covert and overt deterrence coercive measures by the target can stop the adversary from action by adopting a threatening strategy. They vary depending on the goals. Tailored deterrence strategies communicate to different potential adversaries that their aggression would carry unacceptable risks and intolerable costs according to their particular calculations of risk and cost.[166] The idea of cyber deterrence is to provide an overt signal to the alleged opponent that any further attack may be retaliated in kind so that the costs involved will exceed the expected benefits[167] and that counter offensive operations will follow. The robust cyber hygiene and enhanced defence are the most effective deterrence.

As Bruce Schneier points out, the basic techniques for increasing effort, raising risk, and reducing rewards are as true for cyber as for crime prevention: hardening targets, controlling access to facilities, screening exits, detecting offenders, controlling tools, strengthening surveillance, using place managers, reducing peer pressures, and so forth.[168]

Deterrence in cyber response is complicated by the internet anonymity. Nevertheless, Nye states that cyber deterrence is possible even though it can be hard to identify the source of a cyberattack. Attribution problems do not hinder three of the major forms of cyber deterrence: denial (by defence), entanglement, and normative taboos.[169] The fourth form, punishment, is possible only with a proper attribution. However, in most cases, it is more important not to deter the attackers in cyberspace but

---

[164] E. Gartzke and J. Lindsay, Thermonuclear Cyberwar, Research paper, Oxford University Press, Journal of Cybersecurity, 3(1), 2017, 37–48

[165] E. Gartzke and J.R. Lindsay, Weaving tangled webs: offense, defense, and deception in cyberspace. Security Studies, 24: 316–348, 2015, last retrieved on 10 December 2018 at http://deterrence.ucsd.edu/_files/Weaving%20Tangled%20Webs_%20Offense%20 Defense%20and%20Deception%20in%20Cyberspace.pdf

[166] Nuclear posture review, February 2018, Office of the Secretary of Defence, last retrieved on 8 October 2018 at https://media.defense.gov/2018/Feb/02/2001872886/-1/-1/1/2018-NUCLEAR-POSTURE-REVIEW-FINAL-REPORT.PDF

[167] J. Nye, 2017, Deterrence and Dissuasion in cyber space. International security, last retrieved on 10 December 2018 at https://www.belfercenter.org/sites/default/files/files/publication/isec_a_00266.pdf

[168] B. Schneier, Liars and Outliers: Enabling the Trust That Society Needs to Thrive (Indianapolis: John Wiley, 2012), p. 130. Quoted from Joseph S. Nye Jr. 2017, 51 Deterrence and Dissuasion in cyber space. International security, 41 last retrieved on 10 December 2018 at https://www.belfercenter.org/sites/default/files/files/publication/isec_a_00266.pdf

[169] Ibid

to compel them to stop further intrusions. Table 1.2 summarizes some of these major ideas.

Table 1.2: How, Who, and What of Cyber Deterrence and Dissuasion

How	Punishment	Denial/Defence	Entanglement	Norms/Taboos
Who	Both state and non-state actors	Small states and non-states, but not APT	Major states such as China; less so North Korea	Major states; less so rogues; some non-states
What	Major use of force; sanctions against sub-LOAC levels of activity	Some crime and hacking; Imperfect against advanced states	Major use of force; major sub-LOAC actions	LOAC if use of force; taboo on use against civilians; norms against cybercrime

Note: LOAC - Laws of armed conflict
Source[170]: Joseph S. Nye Jr., Deterrence and Dissuasion in cyber space, International Security

Allegedly, there are two areas that define the biggest threat to the nation through cyber space:

Critical information:
- Attacks and intrusions through cyberspace, by either state or organized non-state actors, against information systems to gain strategic information of the national level, and

Critical infrastructure:
- Attacks and intrusions through cyberspace, by either state or organized non-state actors, against critical infrastructure systems to damage or disrupt them.

These definitions can assist policymakers in developing coercion frameworks. They should include the targeted actors against whom a coercion strategy can be directed and who can be either under the direct influence of the state (military, contractors) or require additional efforts and channels of influence (hacktivists, criminals, and terrorists). These strategies can escalate the conflicts between the engaged parties, and spread to other military domains. This depends on the political environment which identifies the proper approach.

---

[170] J.S. Nye, Deterrence and Dissuasion in cyber space. International security, 41, 2017, last retrieved on 10 December 2018 at https://www.belfercenter.org/sites/default/files/files/publication/isec_a_00266.pdf

1.3.5 Attribution of Attacks

Attribution of cyberattacks is one of the most important elements for deterrence and one of major challenges for cyber experts and digital forensics specialists. It has many aspects – technical, strategic, political, legal, and is based on various sources of information – faith-based attribution, assumptions of the target, geopolitical situation, digital forensics and investigation outcomes, which may further reveal (or not) factual information, etc.

The internet architecture and absence of personal identification signs in tools, makes high confidence attribution extremely difficult. In the article "On the Problem of Cyber Attribution" (2017), Berghel comments that "security experts David Clark and Susan Landau provide an overview of the attribution problem from a defensive perspective. They accurately sum up the problem: "Attribution is central to deterrence ... and the Internet was not designed with deterrence in mind. A more technical description is supplied by David Wheeler and Gregory Larsen. Taken together, these papers are notable for the absence of detailed strategies to provide justiciable, evidence-based cyber attribution." Allegedly, today there exists no way to reliably identify the original attacker.[171]

Though it requires a high level of expertise to fully disguise the origin of the cyberattack, we acknowledge that it is possible to disguise attacks, fake a non-intentional accident and identity credentials, copy, change, delete, create digital records, etc. At the same time, it is not less challenging to guarantee the highly accurate attribution.

DARPA Enhanced Attribution Programme[172]

Malicious actors in cyberspace currently operate with little fear of being caught because it is extremely difficult, in some cases perhaps even impossible, to reliably and confidently attribute actions in cyberspace to individuals. The reason cyber attribution is difficult stems at least in part from a lack of end-to-end accountability in the current Internet infrastructure. Cyber campaigns spanning jurisdictions, networks, and devices are only partially observable from the point of view of a defender that operates entirely in a friendly cyber territory (e.g., an organization's enterprise network). The identities of malicious cyber operators are largely obstructed by the use of multiple layers of indirection. The current characterization of malicious cyber campaigns based on indicators of compromise, such as file hashes and command-and-control infrastructure identifiers, allows malicious operators to evade the defenders

---

[171] H. Berghel, 2017 On the Problem of Cyber Attribution, Computer, Vol. 50, Issue 3, last retrieved on 9 December 2018 at https://ieeexplore.ieee.org/stamp/stamp.jsp?tp=&arnumber=7888425 (quoted works of D.D. Clark and S. Landau, "Untangling Attribution," Proc. Workshop Deterring Cyberattacks: Informing Strategies and Developing Options for U.S. Policy, Nat'l Research Council, 2010, pp. 25–40; and D.A. Wheeler and G.N. Larsen, Techniques for Cyber Attack Attribution, IDA Paper P-3792, Inst. for Defense Analyses, Oct. 2003; handle.dtic .mil/100.2/ADA468859.)

[172] I. Crone, Enhanced Attribution, DARPA, last retrieved on 8 December 2018 at https://www.darpa.mil/program/enhanced-attribution

and resume operations simply by superficially changing their tools, as well as aspects of their tactics, techniques, and procedures. The lack of detailed information about the actions and identities of the adversary cyber operators inhibits policymaker considerations and decisions for both cyber and non-cyber response options.

The Enhanced Attribution program aims to make currently opaque malicious cyber adversary actions and individual cyber operator attribution transparent by providing high-fidelity visibility into all aspects of malicious cyber operator actions and to increase the government's ability to publicly reveal the actions of individual malicious cyber operators without damaging sources and methods. The program will develop techniques and tools for generating operationally and tactically relevant information about multiple concurrent independent malicious cyber campaigns, each involving several operators, and the means to share such information with any of a number of interested parties.

The cyber attribution research in the past was mostly focused on the attack source tracking, location of the attacker, or an attacker's intermediary, its technical implementation and how to assign the responsibility.[173,174] However, attribution also includes legal and political aspects – that is the state's decision to publicly announce the attack, accuse another state or condemn the act as unacceptable.

From the technical side, the complexity of the investigation process can be caused by tools providing anonymity (e.g., VPN, Tor, i2p, etc.), proxy channels routing traffic through many intermediary systems, multilayer encryption tunnels, etc. Hence, it is relatively easy to compromise the system and use it as a proxy to attack another target.

Attacks conducted directly from the original system that is directly controlled by the attacker ensure a more stable connection quality, but they do not disguise the origin of the attacks and can be identified and disabled. In many countries, if a serious attack is reported to an Internet Service Provider (ISP) it leads to the shutdown of the connection used by the attacker.

Proxy attacks are routed through one or more intermediary systems. With each additional computer in the proxy chain the stability of the network connection decreases, but the attribution becomes harder, especially if proxy computers are located in different geographical areas.

The contemporary research in attribution encompasses both technical and strategic elements of attribution and multiple models have been developed showing the conditions under which it is rational to tolerate an attack and when it is better to assign blame publicly.[175]

---

[173] David A. Wheeler, Gregory N. Larsen, Techniques for Cyber Attack Attribution, 2003

[174] J. Hunker, C. Gates, M. Bishop, Attribution requirements for next generation Internets, 2011, last retrieved on 5 December 2018 at https://ieeexplore.ieee.org/xpl/mostRecentIssue.jsp?punumber=6095473

[175] B. Edwards et al., Strategic aspects of cyberattack, attribution, and blame, IBM Research, PNAS March 14, 2017 114 (11) 2825-2830; published ahead of print February 27, 2017 last retrieved on 13 December 2018 at https://doi.org/10.1073/pnas.1700442114

The model developed by Herb Lin and presented in "Attribution of Malicious Cyber Incidents"[176], recommends three levels of attribution: machines, human operators, and the ultimate party responsible. Consequently, the attribution should ideally be done on all three levels, namely:

Machines (Technical forensics):

- geolocation and path (GPS coordinates, time zone, language settings),
- tracking a cyber identity (unique identification features of data, IP address, range of the Internet Service Provider, browser, connection tools, aliases and "nicknames")

Human Operator/actual intruder ("Who pressed the button?"):

- initiator identification (behaviour analysis - digital fingerprints, systematic errors, style, etc., login and payment details, etc.).

Adversary:

- Geopolitical situation, interested parties, funding, etc.

The 2017 UNIDIR Report on The United Nations, Cyberspace and International Peace and Security, states that "increasing experience investigating ICT incidents, notably by private sector actors or by cooperation between the latter and law enforcement, has made attribution much more feasible than it was a few years ago. At the same time, regardless of technical progress on reaching attribution findings, actually attributing an incident will likely always be a political decision. In addition, the fact that only a handful of States have advanced attribution capacities and capabilities means that there is limited trust in publicly stated attribution findings, exacerbated by the fact that there are no internationally accepted standards for reaching attribution findings or making attribution claims— issues that will likely remain difficult to resolve in the near term."[177]

Steiger et al. suggest that the open/verbal attribution by the target state's government and third actors, such as international organizations, should be taken more seriously. Even if based on barely sufficient evidence, the open recognition of an attack commits the targeted state's government to (at least) address the attack.[178] Thus, the government fulfils its national obligation to defend the country and its cyberspace, and the international law – any action of self-defence has to be based on

---

[176] H. Lin, "Attribution of Malicious Cyber Incidents: From Soup to Nuts," SSRN Scholarly Paper, Rochester, NY: Social Science Research Network, September 2, 2016

[177] Camino Kavanagh, The United Nations, Cyberspace and International Peace and Security, Responding to Complexity in the 21st Century, UNIDIR, 2017, last retrieved on 9 December 2018 at http://www.unidir.org/files/publications/pdfs/the-united-nations-cyberspace-and-international-peace-and-security-en-691.pdf

[178] S. Steiger et al., Conceptualising conflict in cyberspace, Journal of Cyber Policy, Chatham House, 2018, p. 78

the legitimate recognition of attack. The "actors" public statements have to be triangulated with data from alleged perpetrator, third parties and the expert community, establishing whether and how far an attribution is contested".[179]

The ability to identify, beyond a reasonable doubt, the originator of a cyberattack is essential to enable an effective and legal response. Depending on the level of the threat, the attribution can be more or less essential compared to the fact that the activity is stopped.

In November 2016, Brian J. Egan, the US State Department Legal Adviser, in the remarks prepared for delivery on "International Law and Stability in Cyberspace" stated that "from a legal perspective, the customary international law of state responsibility supplies the standards for attributing acts, including cyber acts, to States. For example, cyber operations conducted by organs of a State or by persons or entities empowered by domestic law to exercise governmental authority are attributable to that State, if such organs, persons, or entities are acting in that capacity. Additionally, cyber operations conducted by non-State actors are attributable to a State under the law of state responsibility when such actors engage in operations pursuant to the State's instructions or under the State's direction or control, or when the State later acknowledges and adopts the operations as its own. [..] Despite the suggestion by some States to the contrary, there is no international legal obligation to reveal evidence on which attribution is based prior to taking appropriate action. There may, of course, be political pressure to do so, and States may choose to reveal such evidence to convince other States to join them in condemnation, for example. But that is a policy choice—it is not compelled by international law".[180]

Draft UN Resolution on Developments in the field of information and telecommunications in the context of international security, 22 October 2018, Seventy-third session, First Committee, A/C.1/73/L.27*[181]

Algeria, Angola, Azerbaijan, Belarus, Bolivia (Plurinational State of), Burundi, Cambodia, China, Cuba, Democratic People's Republic of Korea, Democratic Republic of the Congo, Eritrea, Kazakhstan, Madagascar, Namibia, Nepal, Nicaragua, Pakistan, Russian Federation, Samoa, Sierra Leone, Suriname, Syrian Arab Republic, Tajikistan, Uzbekistan, Venezuela (Bolivarian Republic of) and Zimbabwe

Article 10. States must meet their international obligations regarding internationally wrongful acts attributable to them under international law. However,

---

179 Ibid

180 Brian J. Egan, International Law and Stability in Cyberspace, 35 Berkeley J. Int'l Law. 169 (2017). last retrieved on 15 October 2018 at http://scholarship.law.berkeley.edu/bjil/vol35/iss1/5

181 A/C.1/73/L.27* Developments in the field of information and telecommunications in the context of international security, 22 October 2018, Seventy-third session, First Committee, last retrieved on 18 December 2018 at https://undocs.org/A/C.1/73/L.27

the indication that an information and communications technology activity was launched or otherwise originates from the territory or objects of the information and communications technology infrastructure of a State may be insufficient in itself to attribute the activity to that State. States should note that accusations of organizing and implementing wrongful acts brought against States should be substantiated. In case of information and communications technology incidents. States should consider all relevant information, including the larger context of the event, the challenges of attribution in the information and communications technology environment and the nature and extent of the consequences.

In case of threats to the ruling regime, after the attack detection, proofs collected and attribution identified, the targeted state can engage in negotiations and even appeal to international community. A clear message should be sent to the responsible party that the malicious actions are detected and should be stopped. The compellence message on readiness for countermeasures could be sent and coercive measures could be also announced to increase the deterrence potential of the nation. However, political reasons are taken into consideration in relation to revelation of the adversary, also the country should be ready to reveal some of sensitive information that it might not want to do.

Brian J. Egan, International Law and Stability in Cyberspace,
Remarks prepared for delivery, November 2016
"I want to turn now to the question of what options a victim State might have to respond to malicious cyber activity that falls below the threshold of an armed attack. As an initial matter, a State can always undertake unfriendly acts that are not inconsistent with any of its international obligations in order to influence the behavior of other States. Such acts—which are known as acts of retorsion — may include, for example, the imposition of sanctions or the declaration that a diplomat is persona non grata. In certain circumstances, a State may take action that would otherwise violate international law in response to malicious cyber activity. One example is the use of force in self-defense in response to an actual or imminent armed attack. Another example is that, in exceptional circumstances, a State may be able to avail itself of the plea of necessity, which, subject to certain conditions, might preclude the wrongfulness of an act if the act is the only way for the State to safeguard an essential interest against a grave and imminent peril.

The customary international law doctrine of countermeasures permits a State that is the victim of an internationally wrongful act of another State to take otherwise unlawful measures against the responsible State in order to cause that State to comply with its international obligations, for example, the obligation to cease its internationally wrongful act. Therefore, as a threshold matter, the availability of countermeasures to address malicious cyber activity requires a prior internationally wrongful act that is attributable to another State. As with all countermeasures, this puts the responding State in the position of potentially being held responsible for violating international law if it turns out that there wasn't actually an internationally wrongful act that triggered the right to take countermeasures, or if the responding State made an inaccurate attribution determination. That is one reason why countermeasures should not be engaged in lightly.

> Additionally, under the law of countermeasures, measures undertaken in response to an internationally wrongful act performed in or through cyberspace that is attributable to a State must be directed only at the State responsible for the wrongful act and must meet the principles of necessity and proportionality, including the requirements that a countermeasure must be designed to cause the State to comply with its international obligations—for example, the obligation to cease its internationally wrongful act—and must cease as soon as the offending State begins complying with the obligations in question. The doctrine of countermeasures also generally requires the injured State to call upon the responsible State to comply with its international obligations before a countermeasure may be taken—in other words, the doctrine generally requires what I will call a "prior demand". The sufficiency of a prior demand should be evaluated on a case-by-case basis in light of the particular circumstances of the situation at hand and the purpose of the requirement, which is to give the responsible State notice of the injured State's claim and an opportunity to respond.[182]
>
> *These are the remarks, as prepared for delivery, by Brian J. Egan,*
> *who served as State Department Legal Adviser*
> *from February 22, 2016 to January 20, 2017*

The adequate level of response measures is the next challenging issue. If the desired response is to impose costs on a nation-state ultimately responsible for an intrusion, the conventions and rules of national decision making are relevant, especially those pertaining to making such decisions in a security context.[183]

Lin states that "in the aftermath of a cyberattack, national security decision makers may respond by punishing or retaliating against an adversary's attack through increasing security measures, isolating national telecommunications infrastructure, etc. There are limits on such responses – retaliation or punishment for a hostile act once the act has stopped is prohibited under the UN Charter Section 2(4) if it rises to the level of a use of force. Nevertheless, forceful actions are allowed under Article 51 of the UN Charter if they can be regarded as acts of self-defence in the face of an armed attack, and such actions are often justified as acts of self-defence that deter future attacks. The issue of the cyberattack severity requiring a physical armed attack is still to be identified".[184]

In case of threats to the critical infrastructure, after the attack detection, the security is to be enhanced, proofs collected, and attribution identified. Countermeasures can be employed, negotiations initiated, or a strong message sent overtly or covertly to avoid raising of tension.

In case of intense cyberattacks, like DDoS, security and protection should be enhanced immediately. Counterattacks are not effective at

---

[182] Brian J. Egan, International Law and Stability in Cyberspace, 35 Berkeley J. Int'l Law. 169 (2017). last retrieved on 18 December 2018 at http://scholarship.law.berkeley.edu/bjil/vol35/iss1/5

[183] H. Lin, Attribution of Malicious Cyber Incidents: From Soup to Nut, 2017, last retrieved on 10 December 2018 at https://jia.sipa.columbia.edu/attribution-malicious-cyber-incidents

[184] Ibid

this stage as the network equipment might be overloaded and hence not responsive. Depending on the level of damage and the nation's capacities, an appeal to the international community for help might be voiced. The nation should be ready to demonstrate a certain level of resilience, until international support is provided.

Authoritative attribution of cyberattacks to nation-state actors requires the joint cooperation of technical experts, legal experts and politicians to ensure credibility and due procedural checks, preferably with consideration of the political consequences.

1.3.6 High Impact Cyber Events

Over the years a massive amount of data has been accumulated regarding cyberattacks on the national and international levels.

Valeriano and Maness,[185] using explicit data collection procedures, found that "over an 11-year span, from 2001 to 2011, in the dawn of the potential cyber era, rival states have undertaken 111 total cyber incidents within 45 larger cyber disputes." By incident they understood it to be an isolated cyber operation launched against a state that lasts only a matter of hours, days, or weeks, while a dispute is a longer-term operation that contains several incidents. And nearly half of all cyber incidents in this data can be coded as theft operations, in which a government is attempting to compromise the sensitive information of another government.

In the data set mentioned above, as per 2015, "China is by far the most active state in the use of cyber tactics as a foreign policy tool; it is the most engaged and is the main initiator of cyber conflicts. The United States ranks second, and is also the most targeted state. Other states include rivals in East and South Asia, the Middle East, and the former Soviet Union. Overall, cyber activity does not correlate with power, technology, or resources".[186]

A real-world occurrence that illustrated the dangerous potential of cyberattacks transpired in 2007 when a strike from Israeli forces demolished an alleged nuclear reactor in Syria under construction via a collaborative effort between Syria and North Korea. Accompanied with the strike was a cyberattack on Syria's air defences, which left them unaware of the attack on the nuclear reactor and, ultimately allowed for the attack to occur.[187]

In 2007, a large-scale denial of service attack was launched in Estonia against most of the day to day government services, news portals, banking

---

[185] B. Valeriano and R. Maness, 2015, Cyber War versus Cyber Realities, Cyber Conflict in the International System, Oxford University Press, Oxford, UK, p. 8

[186] Ibid

[187] D.E. Sanger, Syria war stirs new U.S. debate on cyberattacks, 24 February 2014, last retrieved on 8 December 2018 at https://www.nytimes.com/2014/02/25/world/middleeast/obama-worried-about-effects-of-waging-cyberwar-in-syria.html

services, and e-commerce. There is a much speculation on whether or not this was a state directed/sponsored attack, or a spontaneous one. Regardless of attribution, when a state is unable to implement its functions for two weeks it is an issue of the national security.

Estonian Cyber War [188]

As one of the most electronically advanced countries in the world, the Estonian government has increasingly shifted its operations to the virtual domain. Cabinet-level meetings are conducted online and Estonian citizens can cast their votes in national elections via their computers. In 2007, Estonia was ranked 23rd in e-readiness ratings. Almost 61 percent of the population has online access to their bank accounts, and 95 percent of banking transactions are electronic.

In April 2007, the Estonian Parliament, ministries, banks and media institutions were hit by a series of coordinated distributed denial of service (DDoS) attacks. The relocation of the Bronze Soldier of Tallinn, a Soviet-era war memorial, sparked protests and riots among Estonia's Russian minority. These protests were then followed by distributed DDoS attacks to temporarily prevent access to (and in some cases deface) a number of key Estonian websites. A call for action, complete with specific instructions on how to participate in the DDoS attacks, quickly spread through Russian online chat rooms. As a result, the websites of the Ministries of Foreign Affairs and Justice had to shut down, while Prime Minister Andrus Ansip's Reform Party website was defaced. The attack also briefly disabled the national emergency telephone number. Intermittent cyberattacks on national government websites, including the State Chancellery and Federal Electoral Committee, continued well into the middle of May 2007.

Assessments regarding the severity and implications of the attack vary widely. Even though the root of the attack was widely attributed by the community, there was no recourse possible. It is now generally accepted that the attacks were more a "cyber dispute" than a "cyber war" and — while problematic for a small state like Estonia — not particularly sophisticated from a technical standpoint, or especially indicative of what the future might hold. Indeed, Estonia's computer emergency response team (CERT) — as in most other states, a body that coordinates public and private actors to deal with online threats – responded quickly and competently to the attack, using packet filtering and other well-established and successful techniques.

Many have pointed to the events in Estonia as merely a precursor of much more serious and devastating future battles. However, the danger, other commentators argue, is that by overstating the impact of cases such as Estonia, we may drift too far in the other direction, towards a more closed internet, in which the identity of attackers is perhaps easier to ascertain, but in which fundamental freedoms—such as the rights to privacy and freedom of expression—are also less assured. In particular, critics posit that proposals to, for example, "re-engineer the Internet to make attribution, geo-location, intelligence analysis and impact assessment more manageable", as Michael McConnell, the former director of US national intelligence, suggests we should, will also lead to unwanted government control and surveillance over what we write in our emails, type into a search engine, or download from a website.

---

[188] B.S. Buckland, F. Schreier and T.H. Winkler, DCAF HORIZON 2015 working paper No. 1, Democratic Governance Challenges of Cyber Security, p. 26, last retrieved on 15 August 2018 at https://www.dcaf.ch/sites/default/files/publications/documents/OnCyberwarfare-Schreier.pdf

The other cyber conflict illustrated is during the Russia-Georgia war of 2008.

The Georgian Conflict [189]

The cyberattack on Georgia was the first online assault carried out in conjunction with a military offensive. Small-scale DDoS attacks began in June, almost two months before the five-day war between Russia and Georgia over Georgia's breakaway region of South Ossetia.

On the 20th of July, the Shadowserver Foundation, an Internet watchdog group, registered multiple DDoS attacks targeting the official website of Georgian President Mikhail Saakashvili. The attack, which shut down the presidential website for more than 24 hours, was directed by a server in the US, a fact which underlines the borderless nature of the threat.

The DDoS attacks against Georgia's nascent internet infrastructure reached an alarming level on the 8th of August, the first day of the war. On that day, Shadowserver detected the first attack by six different botnets against Georgian government and media websites. As the conflict escalated, so did the online attacks, with Russian hacktivists shutting down and defacing the websites of the President, the Georgian Parliament, the Ministries of Defense and Foreign Affairs, the National Bank of Georgia, and two online news agencies.

The Georgian government reacted swiftly... With Google's permission, the websites of the Ministry of Foreign Affairs and Civil.ge were temporarily transferred to the Blogspot domain where they were better protected against attack. On the 9th of August, the Atlanta-based Internet service provider Tulip Systems Inc., owned by the Georgian-born Nino Doijashvili, began to host the President's website. In a gesture of solidarity, the President of the Republic of Poland, Lech Kaczynski, provided hosting space on his website for the official press releases of the Georgian government. The Estonian government provided substantial assistance by accommodating the website of the Ministry of Foreign Affairs and dispatching two information security specialists to bolster Georgian cyber defences.

According to the Belarusian digital activism expert Evgeny Morozov, coordination of the attacks was largely carried out by an online hacker forum StopGeorgia.ru. Set up hours after the Russian armed forces invaded South Ossetia, this forum featured a constantly updated list of target websites and encouraged visitors to download a free software programme, which allowed them to participate instantly in the attacks.

There is also evidence to suggest that simpler but equally effective SQL injection attacks were also used. These attacks overwhelm the targeted database with millions of junk queries, thereby rendering the corresponding server inoperable. From a hacker's perspective, SQL attacks provide two main advantages. First, when used in combination with traditional DDoS attacks, they are extremely difficult to detect. Second, SQL injection attacks require far fewer computers to achieve the same objectives as DDoS attacks, which cannot be sustained effectively without botnets.

In the end, the cyberattacks inflicted little damage as much of Georgia's economy and critical infrastructure were still not integrated into the Internet. Nonetheless, the campaign waged by nationalist hacktivists, spurred into action with the help of Russian online hacker forums, effectively disrupted the dissemination of information by the Georgian government at a crucial stage in the conflict.

---

[189] Ibid, p. 28

In 2009, the second report from the Information Warfare Monitor based on a 10-month investigation, revealed the existence and operational reach of a malware-based cyber espionage network GhostNet of over 1,295 infected computers collecting intelligence on 103 countries. Up to 30% of the infected computers could be considered high-value diplomatic, political, economic, and military targets, and included the ministries of foreign affairs of Iran, Bangladesh, Latvia, Indonesia, Philippines, Brunei, Barbados and Bhutan; embassies of India, South Korea, Indonesia, Romania, Cyprus, Malta, Thailand, Taiwan, Portugal, Germany and Pakistan; the ASEAN (Association of Southeast Asian Nations) Secretariat, SAARC (South Asian Association for Regional Cooperation), and the Asian Development Bank; news organizations; and an unclassified computer located at NATO headquarters.[190]

In January 2010, Google announced that it had detected a highly sophisticated and targeted attack on its corporate infrastructure originating from China that resulted in the theft of intellectual property from Google. It also stated that 20 other companies were targeted, including the Internet, finance, technology, media and chemical sectors.[191]

In 2010, in Iran, uranium enrichment centrifuges had to be shut down at the Natanz Nuclear Facilities as a result of a sophisticated cyberattack. The sabotage was caused by a newly developed Stuxnet worm, which once hijacked, reprogramed the programmable logic control (PLC) software to give the attached industrial machinery new instructions, while the ICS was showing the stable equipment indicators. It is believed to be the first-known worm designed to target real-world infrastructure such as power stations, water plants and industrial units. It was potentially aiming to disable uranium enrichment infrastructure in Iran.[192]

In December 2014, a cyberattack affected the Hydro and Nuclear Power Co., the nuclear plant operator with 23 of South Korea's nuclear reactors under control. The attack was launched by sending 5,986 phishing emails containing malicious codes to 3,571 employees of the company. The hackers obtained the data belonging to the power plants and demanded the shutdown of three reactors threatening "destruction" in Twitter messages.[193]

---

[190] Information Warfare Monitor (March 2009), 'Tracking GhostNet: Investigating a Cyber Espionage Network, last retrieved on 15 August 2018 at http://www.dhra.mil/perserec/osg/counterintelligence/ cyber-espionage.htm

[191] A new approach to China, Google Official Blog, 12 January 2010, last retrieved on 15 August 2018 at https://googleblog.blogspot.com/2010/01/new-approach-to-china.html

[192] J. Fildes, Stuxnet worm 'targeted high-value Iranian assets', 23 September 2010, last retrieved on 15 August 2018 at http://www.bbc.com/news/technology-11388018

[193] Ju-min Park and Meeyoung Cho, South Korea blames North Korea for December hack on nuclear operator, 17 March 2015, last retrieved on 15 August 2018 at https://www.reuters.com/article/us-nuclear-southkorea-northkorea/south-korea-blames-north-korea-for-december-hack-on-nuclear-operator-idUSKBN0MD0GR20150317

On July 30, 2016, the Russian leading press agencies announced the successful prevention of planned targeted cyberattacks on 20 networks of the Russian federal agencies and defence contractors.[194] Multiple computer systems and networks were infected by the spyware, which after infecting the main computer, started controlling the whole network, capturing traffic, making screenshots, turning on web-cameras and microphones on computers and mobile devices, recording audio and video files, acquiring the role of a full-scale administrator. The affected agencies include federal ministries (Ministry of Defence, Ministry of Economic Development, Ministry of Labour and Social Affairs, Ministry of Emergency), nuclear and space facilities, science and military units, defence industrial facilities.

In December 2015, the Denial of Service in a power plant and multiple substations in Ukraine triggered a power outage. In February 2016, it was acknowledged that BlackEnergy malware was used for the cyberattack.[195]

In 2016, the European Aviation Safety Agency stated that aviation systems are subject to an average of 1,000 attacks each month.[196]

In September 2016, according to a statement of the South Korean politician, 235 gigabytes of classified military documents were stolen from the Republic of Korea Defence Integrated Data Center, including the South Korea–US wartime operational plans to wipe out the North Korean leadership.[197]

In May 2017, the WannaCry and NotPetya ransomware infected (using EternalBlue exploit) over 450 000[198] personal computers and other systems (ATMs, hospital equipment, etc.), that were using a Windows operating system, encrypted stored data and held it for ransom.

In 2018, the city of Leeds, Alabama, paid 8,000 USD in bitcoin, a crypto currency, after their computer systems were taken over. The city was locked out of their systems, from accounting to payroll, and the

---

[194] E. Rozhkov, российские/Federal Security Service stopped a powerful cyber attack on the Russian agencies, 30 July 2016, last retrieved on 15 August 2018 at http://www.vesti.ru/doc.html?id=2782320

[195] US Confirms BlackEnergy malware used in Ukrainian power plant hack, 13 January 2016, last retrieved on 3 August 2018 at http://www.ibtimes.com/us-confirms-blackenergy-malware-used-ukrainian-power-plant-hack-2263008

[196] J. Valero, "Hackers Bombard Aviation Sector with over 1,000 Attacks per Month", Euractiv, 11 July 2016, last retrieved on 15 October 2018 at https://www.euractiv.com/section/justice-homeaffairs/news/hackers-bombard-aviation-sector-withmore-than-1000-attacks-per-month/

[197] Christine Kim, North Korea hackers stole South Korea-U.S. military plans to wipe out North Korea leadership: lawmaker, 10 October 2017, last retrieved on 15 August 2018, at https://www.reuters.com/article/us-northkorea-cybercrime-southkorea/north-korea-hackers-stole-south-korea-u-s-military-plans-to-wipe-out-north-korea-leadership-lawmaker-idUSKBN1CF1WT

[198] WannaCry coding mistakes can help files recovery even after infection, 2 June 2017, last retrieved on 2 August 2018 at http://thehackernews.com/2017/06/wannacry-ransomware-unlock-files.html

department heads lost access to files and email and were no longer able to pay city employees or bills electronically, instead they had to resort to the use of handwriting and delivering checks.[199]

In February 2018, the two largest in history DDoS attacks helped cybercriminals launch nearly 15,000 cyberattacks against 7,131 unique targets in ten days. The targets included online services and websites such as Google, Amazon, QQ.com, 360.com, PlayStation, OVH Hosting, VirusTotal, Comodo, GitHub (1.35 Tbps attack), Royal Bank, Minecraft and RockStar games, Avast, Kaspersky, Epoch Times newspaper, and Pinterest.[200]

In July 2018, a Ukrainian intel agency claimed it stopped a cyberattack against a chlorine plant that was launched using the notorious VPNFilter malware.[201] The attack was allegedly geared at disrupting the stable operation of the plant, which provides NaClO (sodium hypochlorite, aka liquid chlorine) for water treatment.

1.3.7 Organised Cybercrime

Cyberattacks are the fastest growing crime
and predicted to cost the world 6 trillion US dollars annually by 2021[202]

Cybercrime appeared together with the cyberspace, acquired transnational dimension and has now reached the level of the global "cybercrime epidemics". The 2010 UN General Assembly resolution on cybersecurity addressed cybercrime as one of the major global challenges.[203]

Cybercrime growth and evolution, was initiated by the physical damage and property theft of computer systems and stored data in the 1960s, passed through data manipulation, theft of intellectual property, personal and financial data, fraud in transactions, raised to boundless transnational level and is currently represented by highly sophisticated

---

[199] Michael Clark, Leeds city government recovering after computer system held hostage, 1 March 2018, last retrieved on 15 August 2018, https://www.cbs42.com/news/local/leeds-city-government-recovering-after-computer-system-held-hostage/1002283043

[200] Swati Khandelwal, Over 15,000 Memcached DDoS Attacks Hit 7,100 Sites in Last 10 Days, 9 March 2018, at https://thehackernews.com/2018/03/memcached-ddos-attack.html

[201] Ukrain claims it blocked VPNFilter attack at chemical plant, 13 July 2018, last retrieved on 13 August 2018 at https://www.theregister.co.uk/2018/07/13/ukraine_vpnfilter_attack/

[202] Cyberattacks are the fastest growing crime and predicted to cost the world $6 trillion annually by 2021, 2019 Official Annual Cybercrime Report Announced By Cybersecurity Ventures, 13 December 2018, last retrieved on 28 December 2018 at https://www.prnewswire.com/news-releases/cyberattacks-are-the-fastest-growing-crime-and-predicted-to-cost-the-world-6-trillion-annually-by-2021-300765090.html

[203] UNGA Resolution 64/211: Creation of a global culture of cybersecurity and taking stock of national efforts to protect critical information infrastructures, A/RES/64/211, 2010, last retrieved on 21 May 2018 at at http://www.un.org/en/ga/search/view_doc.asp?symbol=A/RES/64/211

crimes such as social engineering techniques, advanced persistent threats, injection of malware implants into the supply chain, use of zero-days, artificial intelligence and deep learning.

Cyber criminals vary, from the lowest level of individuals with minimal skills, to highly skilled cyber criminal groups sharing global profits and not bound by any rules. Although many of the different types of attackers can clearly be considered cyber criminals, those that participate in organised crime can be considered to be in a different category entirely. Those involved in the efforts of organized crime use malware, DDoS attacks, identity theft, phishing, ransomware-as-a-service, stolen credit card databases, and any number of other tactics that might be the means to the particular end they wish to accomplish.[204]

The report on the 2017 NSA data leak published in March 2018[205] uncovered government tools to track and monitor offensive cyber operations of cyber criminals and other nations. This event has proven that cybercrime carries out cyber offensive operations so large, that it requires tools and monitoring capabilities of that magnitude.

According to a Lloyd's of London report,[206] a major cyberattack on a cloud services provider such as Amazon could trigger economic losses of up to 53 billion USD, a figure on par with a catastrophic natural disaster such as Hurricane Sandy, which hit much of the eastern United States in 2012.

The national cybercrime statistics, though very diverse and depending on recording practices, legislations, selected respondents, etc., is very informative and useful to show the trends and draw attention to the growing number of cybercrimes.

The 2019 Annual Cybercrime Report, announced in December 2018, states that cybercrime will cost the world 6 trillion USD annually by 2021, up from 3 trillion USD in 2015. This represents the greatest transfer of economic wealth in history, risks the incentives for innovation and investment, and will be more profitable than the global trade of all major illegal drugs combined.[207]

---

[204] Understanding cybercrime: Phenomena, challenged and legal response, ITU report, 2014

[205] K. Zetter, Leaked Files Show How the NSA Tracks Other Countries' Hackers, 7 March 2018, last retrieved on 17 October 2018 at https://theintercept.com/2018/03/06/leaked-files-show-how-nsa-tracks-other-countries-hackers/

[206] Extreme cyber-attack could cost as much as Superstorm Sandy, 17 July 2017, last retrieved on 23 July 2018 at https://www.lloyds.com/news-and-risk-insight/press-releases/2017/07/cyber-attack-report

[207] Cyberattacks are the fastest growing crime and predicted to cost the world $6 trillion annually by 2021, Press release for the 2019 Official Annual Cybercrime Report announced by Cybersecurity Ventures, 13 December 2018, last retrieved on 15 December 2018 at https://www.prnewswire.com/news-releases/cyberattacks-are-the-fastest-growing-crime-and-predicted-to-cost-the-world-6-trillion-annually-by-2021-300765090.html

In 2017, the Ponemon Institute study evaluated the responses of 2,182 interviews from 254 companies in seven countries—Australia, France, Germany, Italy, Japan, United Kingdom and the United States. The study examined the total costs organizations incur when responding to cybercrime incidents, including the costs to detect, recover, investigate and manage the incident response, the costs that result in after-the-fact activities and efforts to contain additional costs from business disruption and the loss of customers.

Figure 1.4 presents the estimated average cost of cybercrime for the evaluated seven countries for the past three years. Companies in the United States report the highest total average cost at 21 million US dollars and Australia reports the lowest total average cost at 5.41 million US dollars.

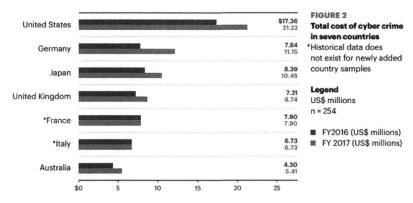

Figure 1.4: Total cost of Cyber Crime in seven countries

Source: The 2017 cost of Cybercrime study, Ponemon Institute LLC

Cyber breaches recorded by businesses have almost doubled in five years, from 68 per business in 2012 to 130 per business in 2017.[208] "Having been choked off by law enforcement successes in 2010–2012, 'dark net' markets for malware goods and services have seen a resurgence:[209] in 2016 alone, 357 million new malware variants were released and 'banking trojans' designed to steal account login details could be purchased for as little as 500 USD."[210]

---

[208] Cost of Cyber Crime Study, Accenture, 2017, last retrieved on 15 October 2018 at https://www.accenture.com/t20170926T072837Z__w__/us-en/_acnmedia/PDF-61/Accenture-2017-CostCyberCrimeStudy.pdf

[209] L. Kessem, Commercial Malware Makes a Comeback in 2016, IBM Security Intelligence, 29 March 2017, last retrieved on 15 December 2018 at https://securityintelligence.com/commercial-malware-makes-a-comeback-in-2016/

[210] Internet Security Threat Report, Volume 22, Symantec, April 2017, last retrieved on 14 October 2018 at https://www.symantec.com/content/dam/symantec/docs/reports/istr-22-2017-en.pdf

The crime techniques evolve. Among the newest emerging threats is the use of the consumers devices for cryptocurrency mining instead of credential theft or ransomware spread. Symantec announced that the coin mining gold rush resulted in an 8,500 percent increase in detections of coin miners on endpoint computers in 2017.[211] This indicates that many cybercriminals are interested in using a victim's computer power and resources to mine cryptocurrencies instead of stealing any personal data or money.[212]

In addition, cybercriminals have an exponentially increasing number of potential targets, as the use of cloud services continues to accelerate and the Internet of Things is expected to expand from approximate 8.4 billion devices in 2017 to a projected 20.4 billion in 2020.[213]

Previously large-scale cyberattacks would be considered average. For example, in 2016, companies revealed breaches of more than 4 billion data records, more than the combined total for the previous two years. Distributed denial of service attacks using 100 gigabits per second (Gbps) were once exceptional but have now become commonplace, jumping in frequency by 140% in 2016 alone. And attackers have become more persistent – in 2017 the average DDoS target was likely to be hit 32 times over a three-month period.[214]

A 2018 study, commissioned by Bromium, Inc., examined how new criminality platforms and a growing cybercrime economy resulted in 1.5 trillion US dollars in illicit profits being acquired, laundered, spent and reinvested by cybercriminals annually. The research was one of the first to view the dynamics of cybercrime through the revenue flow and profit distribution. This 1.5 trillion USD figure includes: 860 billion – Illicit/illegal online markets; 500 billion – theft of trade secrets/IP; 160 billion – data trading; 1.6 billion – crimeware-as-a-service; 1 billion – ransomware.[215]

---

[211] Internet Security Threat Report, Volume 23, Symantec, March 2018, last retrieved on 14 October 2018 at http://images.mktgassets.symantec.com/Web/Symantec/%7B3a70beb8-c55d-4516-98ed-1d0818a42661%7D_ISTR23_Main-FINAL-APR10.pdf?aid=elq_

[212] A. O'Driscoll, 100+ Terrifying Cybercrime and Cybersecurity Statistics & Trends [2018 EDITION], 2 October 2018, last retrieved on 28 October 2018 at https://www.comparitech.com/vpn/cybersecurity-cyber-crime-statistics-facts-trends/#gref

[213] Gartner, "Gartner Says 8.4 Billion Connected 'Things' Will Be in Use in 2017", 7 February 2017, last retrieved on 14 August 2018 at https://www.gartner.com/en/newsroom/press-releases/2017-02-07-gartner-says-8-billion-connected-things-will-be-in-use-in-2017-up-31-percent-from-2016

[214] Q2 2017 State of the Internet/Security Report, Akamai, 2017, last retrieved on 18 October 2018 at https://www.akamai.com/de/de/multimedia/documents/state-of-the-internet/q2-2017-state-of-the-internet-security-report.pdf

[215] Hyper-connected web of profit emerges, as global cybercriminal revenues hit $1.5 trillion annually, Press release, 20 April 2018, last retrieved on 28 August 2018 at https://www.bromium.com/company/hyper-connected-web-of-profit-emerges-as-global-cybercriminal-revenues-hit-1-5-trillion-annually/

This report finds evidence that cybercrime revenues often exceed those of legitimate companies – especially at the small to medium enterprise size. Revenue generation takes place at a variety of levels – from large 'multi-national' operations that can make profits of over 1 billion USD; to smaller operations where profits of 30,000–50,000 USD are the norm. Individual hackers may only earn around 30,000 USD per year. Managers can earn up to 2 million USD per job – often with just 50 stolen card details at their disposal.[216]

As in the legitimate economy, criminal enterprises are going through digital transformation and diversifying into new areas of crime. The research by Bromium revealed that cybercriminals were found to be reinvesting 20 percent of their revenues into further crime, which suggests up to 300 billion USD is being used to fund future cybercrime and other serious types of crime – including drug manufacturing, human trafficking or terrorism. For example, the takedown of Alphabay – one of the largest dark web online markets – revealed that in addition to more than 250,000 listings for illegal drugs, there were also listings for toxic chemicals, firearms, counterfeit goods, malware, and over 100,000 listings for stolen and fraudulent identification documents and access devices. This demonstrates that platform criminality can easily adapt to include other areas of crime.[217]

As per the estimates, cybercriminals are going to create 3.5 million new unfilled cybersecurity jobs by 2021.[218] Compare that with the one million openings in 2016, this is an increase of 350 percent in five years. With that increase comes serious cybersecurity revenue dedication. Businesses are investing a remarkable amount of money into hiring security professionals, maintaining customer privacy and avoiding ransomware attacks. In 2017 alone, all those defence measures cost businesses 86.4 billion USD.[219]

The issue of cybercrime is on the global UN agenda. Thus, in December 2018, the UN General Assembly adopted, by majority vote, the resolution on Countering the Use of Information and Communications Technologies for Criminal Purposes. The resolution was supported by 94 Member States with 59 voting against and 33 abstaining. The key aim of the initiative was to launch a broad transparent political discussion on combating information crime and to search for and create responses to one of today's most pressing challenges. The resolution is aimed at promoting

---

[216] Ibid

[217] Ibid

[218] S. Morgan, Cybersecurity labor crunch to hit 3.5 million unfilled jobs by 2021, 6 June 2017, last retrieved on 22 April 2018 at https://www.csoonline.com/article/3200024/security/cybersecurity-labor-crunch-to-hit-35-million-unfilled-jobs-by-2021.html

[219] Cybersecurity Challenges and Trends: What to Expect in 2018, 10 January 2018, last retrieved on 15 May 2018 at https://www.globalsign.com/en/blog/cybersecurity-trends-and-challenges-2018/

a global consensus and working out concrete and practical approaches to countering cybercrime in the absence of effective international legal instruments.[220] The resolution was co-authored by 36 countries, including all BRICS, SCO and CSTO partners. The majority of the developing counties in Asia, Africa and Latin America favoured adoption.

Even as preliminary observations, all of the abovementioned explicitly shows the rising danger and threatening consequences of the cyber criminal activities. Cybercrime will only increase following the sophistication and accessibility of technology and the increase in the number of Internet users. Comprehensive cyber anti-crime measures should be mandatory and include technical protection, policies, legal instruments, training of professionals and the education of the wider public. Possible training activities can range from age-specific crime prevention training in schools to real-time cyber incident simulations for top government officials to help them to prevent attacks.[221]

---

[220] Press release on the UN General Assembly adoption of a Russian-proposed resolution on combating cyber crime, 18 December 2018, last retrieved on 22 December 2018 at http://www.mid.ru/mezdunarodnaa-informacionnaa-bezopasnost/-/asset_publisher/UsCUTiw2pO53/content/id/3449030

[221] Understanding cybercrime: Phenomena, Challenges and Legal response, Report, ITU, November 2014, last retrieved on 17 October 2018 at https://www.itu.int/en/ITU-D/Cybersecurity/Documents/cybercrime2014.pdf

2 Cyber Defence and Countermeasures

Cyber defence is aimed at protecting cyberspace from external and internal threats. It is a computer network-led defence mechanism which includes responses to actions and critical infrastructure protection, information assurance for organizations, government entities and other possible networks. It focuses on preventing, detecting and providing timely responses to attacks or threats so that no infrastructure or information is tampered with. With the growth in volume as well as the complexity of cyberattacks, cyber defence is essential in order to protect sensitive information as well as to safeguard assets.[1]

There exist a number of terms related to the cyber defence area: computer security, network security, IT security, cyber security, information security. Though all of them are aimed at protecting internal and external cyber space, each one has its own additional meaning. And the debates on their proper interpretation and application are on-going.

The West, including the NATO allies, see cyber security as the security of computer and information systems as physical and logical entities, and information assurance or information security as referring to security of the content.[2] Russia, China, and several other states use the broader term "information security". The Doctrine of Information Security of the Russian Federation of 5 December 2016, defines the information security as "the state of protection of the individual, society and the State against internal and external information threats, allowing to ensure the constitutional human and civil rights and freedoms, the decent quality and standard of living for citizens, the sovereignty, the territorial integrity and sustainable socio-economic development of the Russian Federation, as well as defence and security of the State."[3] The term "cyber" is used as a part of the information security, while the term "cyberspace" covers the connected devices and their use.

---

[1] Definition "Cyber defense", Technopedia, last retrieved on 18 October 2018 at https://www.techopedia.com/definition/6705/cyber-defense
[2] M. Raud, China and Cyber: Attitudes, Strategies, Organization, The NATO Cooperative Cyber Defence Centre of Excellence, Tallin, 2016, last retrieved on 2 August 2018 at https://ccdcoe.org/sites/default/files/multimedia/pdf/CS_organisation_CHINA_092016_FINAL.pdf
[3] The Doctrine of Information Security of the Russian Federation of 5 December 2016, approved by the Decree of the President of the Russian Federation No 646 of December 5, 2016, last retrieved on 17 December 2018 at http://www.mid.ru/en/foreign_policy/official_documents/-/asset_publisher/CptICkB6BZ29/content/id/2563163

Basic notions related to international information security include:
(a) **Information area**. The sphere of activity involving the creation, transformation or use of information, including individual and social consciousness, the information and telecommunications infrastructure and information itself;
(b) **Information resources**. Information infrastructure (hardware and systems for creating, processing, storing and transmitting information), including data files and bases, and information and information flows;
(c) **Information war**. Confrontation between States in the information field, with a view towards damaging information systems, processes and resources and vital structures, and undermining another State's political and social systems, as well as the mass psychological manipulation of a State's population and the destabilization of society;
(d) **Information weapon**. Means and methods used with a view to damaging another State's information resources, processes and systems; use of information towards the detriment of a State's defence, administrative, political, social, economic or other vital systems, and the mass manipulation of a State's population with a view to destabilizing society and the State;
(e) **Information security**. Protection of the basic interests of the individual, society and the State in the information area, including the information and telecommunications infrastructure and information per se with respect to its characteristics, such as integrity, objectivity, accessibility and confidentiality;
(f) **Threat to information security**. Factors that endanger the basic interests of the individual, society and the State in the information area;
(g) **International information security**. The state of international relations that excludes the violation of international stability and the creation of a threat to the security of States and the international community in the information era; ...
(k) **International information terrorism**. The use of telecommunications or information systems and resources, or the influencing of those systems or resources, in the international information area for terrorist purposes;
(l) **International information crime**. The use of telecommunications or information systems and resources, or the influencing of those systems or resources, in the international information area for illegal purposes.

Report of the Secretary-General, UNGA Fifty-fourth session, A/54/213[4]
"Developments in the field of information and telecommunications in the context of international security" (Reply from Russia) 10 August 1999

Information is a weapon in itself and one of the major targets in any war. Over the course of history the value of information has only increased and the means of getting it have been further perfected. Information, if successfully gathered, can now provide the full spectrum of data about the target, both material and virtual, including their dreams and hopes.

---

[4] Developments in the field of information and telecommunications in the context of international security. Report of the Secretary-General, UNGA Fifty-fourth session, 10 August 1999, https://undocs.org/A/54/213

2.1 Information Security

Information security is a discipline that is aimed to protect information throughout its life cycle, and information systems, its containers, from unauthorized access and any manipulations. It protects the information space as a whole, including the content influencing human minds and behaviour; and it applies specific defence mechanisms at storage, procession and transmission.

As per the US Field Manual for Information Operations,[5] the Army's foundational doctrine for information operations, the information environment is comprised of three dimensions: physical, informational, and cognitive. The physical dimension of the information environment covers connective infrastructure to support the transmission, reception, and storage of information. The informational dimension is the content or data itself, it also refers to the flow of information, either through text or images, that can be collected, processed, stored, disseminated, and displayed. It links the physical and cognitive dimensions, with the latter reflecting the minds of those who are affected by and act upon information. Cognitive dimension focuses on the societal, cultural, religious, and historical contexts that influence the perceptions of those producing the information and of the targets and audiences receiving the information. This explains why some governments (e.g., China) apply a control-seeking and restrictive approach to information security.[6]

Information has become a super power and is used to build opinions and manipulate selected target groups and societies as a whole. Corporations Alphabet (Google), Amazon, Apple, Facebook and Microsoft—are the five most valuable listed commercial organizations in the world.[7] Governments, the military, critical infrastructure services, and CBRNe industries have very strict laws and regulations regarding how information is handled and controlled, thus protecting civilian as well as strategic data.

---

[5] FM 3-13 Information Operations, December 2016, Headquarters Department of the Army Washington, DC, last retrieved on 27 August 2018 at https://www.globalsecurity.org/military/library/policy/army/fm/3-13/fm3-13_2016.pdf

[6] M. Raud, China and Cyber: Attitudes, Strategies, Organization, The NATO Cooperative Cyber Defence Centre of Excellence, Tallin, 2016, last retrieved on 15 August 2018 at https://ccdcoe.org/sites/default/files/multimedia/pdf/CS_organization_CHINA_092016_FINAL.pdf

[7] The world's most valuable resource is no longer oil, but data, 6 May 2017, last retrieved on 17 October 2018 at https://www.economist.com/leaders/2017/05/06/the-worlds-most-valuable-resource-is-no-longer-oil-but-data

> **German politicians targeted in mass data attack, 4 January 2019**[8]
> Hundreds of German politicians, including Chancellor Angela Merkel, have had personal details stolen and published online.
> Contacts, private chats and financial details were put out on Twitter that belong to figures from every political party except the far-right AfD.
> Data from celebrities and journalists were also leaked.
> It is unclear who was behind the attack, which emerged on Twitter in the style of an advent calendar last month.

2.1.1 Information Security as a Field of Study

Information protection appeared alongside information itself. Currently, information security is a well-grounded and, at the same time, dynamically evolving area of services that adjusts accordingly to the development of information types, its containers and channels of distribution. It protects confidentiality, integrity and the availability of information assets, whether in storage, processing, or transit.[9] It is achieved via the application of security policies in organizations, education and training of staff, and enhancement of technology.

Information security is a multidisciplinary area of study and professional activity which develops and implements security mechanisms of all available types (technical, organizational, human-oriented and legal) in order to keep information in all its locations (within and outside the organization's perimeter) and, consequently, information systems, where information is created, processed, stored, transmitted and destroyed, free from threats.[10]

In this book we will cover a few issues related to information security management and the related environment security (computer and network security).

Initially, information security consisted of physical security and document classification. The primary threats to security were physical theft of equipment, espionage against the products of the systems, and sabotage. With the development of networks and online connections to accomplish more complex and sophisticated tasks, it became possible to transfer data via new communication channels and necessary to protect it from the ICT vulnerabilities.

---

[8] German politicians targeted in mass data attack, 4 January 2019, last retrieved on 4 January 2019 at https://www.bbc.co.uk/news/world-europe-46757009

[9] A. Singh, A. Vaish and P.K. Keserwani, Information Security: Components and Techniques, Volume 4, Issue 1, January 2014, last retrieved on 18 October 2018 at http://ijarcsse.com/Before_August_2017/docs/papers/Volume_4/1_January2014/V4I1-0528.pdf

[10] F. Almeida and I. Portela (eds.), Organizational, Legal, and Technological Dimensions of IS Administrator, IGI Global Publishing, September, 2013

In 1966 concerns relating to the protection of private data in the US had become so alarming that the American Congress dedicated three days to the hearings on "The Computer and Invasion of Privacy". There was also a growing realization within the defence sector that vulnerabilities in networked computers could (and would) be exploited by hackers to commit fraud and steal or manipulate information.[11] The theft of sensitive information, including personal data, has grown exponentially since then.

On 20 July 2018, in Singapore, cybercriminals stole personal data belonging to 1.5 million people (about a quarter of the population), through a "deliberate, targeted and well-planned" attack, according to a government statement.[12] The data of Prime Minister Lee Hsien Loong, including information on his medicines, was "specifically and repeatedly targeted".

"Health records are often targeted as they contain valuable information to governments," says Eric Hoh, the Asia Pacific president of security company FireEye. "Nation states increasingly collect intelligence through cyber espionage operations which exploit the very technology we rely upon in our daily lives," he says, adding: "Many businesses and governments in South East Asia face cyber threats, but few recognize the scale of the risks they pose."[13]

At the core of information security is information assurance, the act of maintaining the Confidentiality, Integrity and Availability (CIA) of information, ensuring that information is not compromised in any way when critical issues arise. These issues include natural disasters, networks and computers, etc.[14]

Information assurance[15] is a multidisciplinary area of study and professional activity which aims to protect businesses by reducing risks associated with information and information systems by means of a comprehensive and systematic management of security countermeasures, which is driven by risk analysis and cost-effectiveness.

The famous triangle model of Confidentiality, Integrity and Availability (CIA) of data reflects information security goals in general. Data confidentiality ensures the access to information for authorized persons only, its integrity refers to preventing unauthorized modifications

---

[11] Camino Kavanagh, The United Nations, Cyberspace and International Peace and Security, Responding to Complexity in the 21st Century, UNIDIR, 2017, last retrieved on 9 December 2018 at http://www.unidir.org/files/publications/pdfs/the-united-nations-cyberspace-and-international-peace-and-security-en-691.pdf

[12] Singapore personal data hack hits 1.5m, health authority says, 20 July 2018, last retrieved on 2 August 2018 at https://www.bbc.co.uk/news/world-asia-44900507

[13] Ibid

[14] S. Samonas and D. Coss, 2014. "The CIA Strikes Back: Redefining Confidentiality, Integrity and Availability in Security". Journal of Information System Security. 10 (3): 21–45.

[15] F. Almeida and I. Portela (eds.), Organizational, Legal, and Technological Dimensions of IS Administrator. IGI Global Publishing, September, 2013

to data or system functions, and the availability of data which also means its accessibility when needed.

> **Confidentiality**
> Examples of the violation of confidentiality may include laptop theft, password theft, or sensitive emails being sent to the incorrect individuals. The violator can be either the owner of the information (by accident), or the malicious attack (by intent).
> **Integrity**
> Examples of the integrity breach can be an opened letter in the mail, or, a Man-in-the-Middle attack in a cybersecurity domain. Similar to the measures used to provide confidentiality, encryption can be used to help provide integrity by preventing data manipulation without proper authorization. In particular, hashes or message digests, such as MD5 and SHA1, are often used to ensure that messages or files have not been altered from the original by creating a fingerprint of the original data that can be tracked over time.
> **Availability**
> In the cyber space, the Denial-of-Service (DoS) attack is an example of the compromised availability of information. During DoS the users cannot access the service, and thus it is deemed unavailable.

However, with technological advancements, the CIA model can no longer fully address the changed environment, with evolved threats which include accidental or intentional damage, destruction, theft, unintended or unauthorized modification, or other human or nonhuman misuse. The expanded model varies with additional critical characteristics of information depending on the authors' approaches, but in general covers confidentiality, integrity, availability, privacy, authenticity and trustworthiness, non-repudiation, accountability and auditability.

> **Nonrepudiation** ensures that a transferred message has been sent and received by the parties who have actually sent and received the message, and guarantees that both of them cannot deny this process.
>   Nonrepudiation is a method of guaranteeing message transmission between parties via digital signature and/or encryption. It is often used for digital contracts, signatures and email messages.
>   While nonrepudiation is a worthy electronic security measure, professionals in this arena caution that it may not be 100 percent effective. Phishing or man-in-the-middle (MITM) attacks can compromise data integrity. In addition, it is important to note that a digital signature is the same whether it is authentic or faked by someone who has the private key.
>   This problem has been countered by the U.S. Department of Defense with the development of the common access card, a type of smart card designed for active duty military personnel, civilian personnel, the National Guard and others that are privy to confidential defense information.[16]

---

[16]   Definition—Nonrepudiation, Techopedia, last retrieved on 16 October 2018 at https://www.techopedia.com/definition/4031/nonrepudiation

Not all information is equal, thus its protection and risk management measures vary harmonizing expected security level with expenditures, and defining appropriate procedures and protection requirements. Information is classified with the consideration of its value to the organization, its life cycle period, laws and other regulatory requirements. The Information Systems Audit and Control Association[17] (ISACA) and its Business Model for Information Security also serve as a tool for security professionals to examine information security from a systems perspective, creating an environment where security can be managed holistically, allowing actual risks to be addressed.

Table 2.1: Examples of Information Security Classification

Business sector[18]	Government sector
• Public • Private • Sensitive • Confidential	• Unclassified • Unofficial • Official • Classified/Confidential • Protected • Secret • Top Secret

Information security plan is developed and adopted by each organization to establish security procedures and employ the proper tools for evaluation, monitoring and reporting of any potential breach of the established protection level.

Information classification starts with the appointment of senior staff responsible for classification of information, they then classify and label information. Afterwards, information security policy is applied for each level of information, security controls assigned for each classification, and responsible personnel appointed for clearance and trained.

In 1992, and as revised in 2002, the Organization for Economic Cooperation and Development (OECD) Guidelines for the Security of Information Systems and Networks[19] proposed nine generally accepted principles: awareness, responsibility, response, ethics, democracy, risk assessment, security design and implementation, security management, and reassessment. In 2001, the National Institute of Standards and

---

[17] ISACA, last retrieved on 20 October 2018 at https://www.isaca.org/pages/default.aspx

[18] What are the data classifications, last retrieved on 8 December 2018 at https://security. uwo.ca/information_governance/standards/data_classification/data_classifications. html

[19] OECD Guidelines for the Security of Information Systems and Networks, 25 July 2002, last retrieved on 14 October 2018 at https://www.oecd.org/sti/ieconomy/15582260.pdf (link in French)

Technology (NIST) proposed a more detailed and technical collection of 33 Engineering Principles for Information Technology Security.[20]

> In December 2018, almost 1,000 North Korean defectors had their personal data leaked after a computer at a South Korean resettlement centre was hacked. The hackers' identity and the origin of the cyber-attack are not yet confirmed.[21]

Over the period of history of information security in cyberspace, multiple levels of security were implemented to protect critical computer systems and maintain the integrity of their data, e.g., access to sensitive strategic locations is controlled by means of badges, keys, facial recognition. The growing need to maintain national security eventually led to more complex and more technologically sophisticated security measures, with cryptographic mechanism as one of them.

2.1.2 Cryptography

Cryptography is a key information security technology which hides the meaning of the message through transforming its content to an unintelligible form, which is then unclear to unauthorized users. This transformation process is called encryption and it is based on complex mathematics and computer science methods. The encrypted information can be decrypted, i.e. transformed by the cryptographic key back into its original usable form.

Cryptography protects information whether it is in storage or in transit (either electronically or physically) and it helps to achieve several goals of information security, i.e. confidentiality (secrecy), integrity, and authentication. It protects confidentiality, as the encrypted information cannot be used by an unauthorized person without the proper keys for decryption; it ensures the integrity (or accuracy) of information through the use of hashing algorithms and message digests; it enhances authentication (and non-repudiation) services through digital signatures, digital certificates, interactive proofs, secure computation or a Public Key Infrastructure (PKI).

Two major techniques used in encryption are symmetric and asymmetric encryption. In symmetric encryption, one key is used to encrypt and decrypt the message. The decryption procedure is faster as the key is comparatively short, but it requires better security for the key itself. Asymmetric technique uses a public key to encrypt a message and a private key to decrypt it. Both systems have their strengths and weaknesses and

---

[20] Engineering Principles for Information Technology Security, NIST, 1 July 2001, last retrieved on 27 July 2018 at https://www.nist.gov/publications/engineering-principles-information-technology-security

[21] North Korea defector hack: Personal data of almost 1,000 leaked, 28 December 2018, last retrieved on 30 December 2018 at https://www.bbc.co.uk/news/world-asia-46698646

the most appropriate solution should be identified based on the computer potential, type of data and access, key management strategies, costs and available funding, etc. A strong key management process is essential to prevent unauthorized access to sensitive information, as the key theft is one of the most wide-spread methods of attacking a cryptosystem.

Modern cryptographic solutions are very complex and applied based on the approved standards that have undergone rigorous checks by independent experts.

Selected symmetric systems are presented by the Data Encryption Standard (DES), Triple DES, Advanced Encryption Standard (AES), International Data Encryption Algorithm (IDEA), Blowfish, Twofish, RC4, Skipjack, etc.

The Data Encryption Standard (DES) was developed by the US in the mid-1970s and selected by the National Bureau of Standards as an official US Federal Information Processing Standard. It was also used internationally. In the late 1990s it was replaced by the Triple DES, which encrypted the plaintext in three iterations. It is still widely used, e.g., in ATM encryption, private e-mails, etc.[22] It remained in effect until 2002, when it was superseded by the Advanced Encryption Standard (AES), recognized by the US government as suitable to protect classified information, and which is currently one of the most popular forms of encryption standards.

Use of asymmetric systems enhances information security through the use of digital signatures and stronger key management. The Ron Rivest-Shamir-Adleman's encryption algorithm (RSA), which was publicly described in 1977, was a big step forward in the public-key cryptography. It is currently widely used in electronic commerce protocols and is built into the operating systems of Microsoft, Apple, and Linux. Other examples include ElGamal encryption, Cramer-Shoup cryptosystem (1998), and the Elliptic Curve Cryptography (ECC), applied from 2004 onwards using a smaller key size, thus decreasing transmission requirements.

Public key infrastructure (PKI) is more secure and more computationally expensive than symmetric encryption. PKI is used for secret key exchange, to initiate the symmetric encryption. There also exists a hybrid encryption, where the key is provided and at the same time - a symmetric key cipher is also provided. Nonrepudiation is usually provided through digital signatures, and a hash function.

Interim solutions, e.g., Signcryption, includes combinations of the digital signature and the public key encryption in a single logical step, ensuring authentication and confidentiality simultaneously with a cost lower than signing and encrypting the message independently. The ring signcryption added anonymity through enabling the user to signcrypt a

---

[22] D. Dunning, What encryption is used on an ATM machine? Last retrieved 15 July 2018 at https://itstillworks.com/encryption-used-atm-machine-11369995.html

message along with the identities of a set of potential senders, including himself, without revealing which user in the set has actually produced the signcryption.[23]

Cryptographic hash functions are a type of cryptographic algorithm, which are used to verify the authenticity of data retrieved from an untrusted source or to add a layer of security. They take a message of any length as input, and output a short, fixed length hash, which can be used in (for example) a digital signature. The most widely used cryptographic hash functions are MD4, MD5 (Message Digest), and SHA/SHS (Secure Hash Algorithm or Standard).

Message authentication codes are much like cryptographic hash functions, except that a secret key can be used to authenticate the hash value upon receipt; this additional complication blocks an attack scheme against bare digest algorithms, and so has been thought worth the effort.

Encryption and signature schemes are fundamental cryptographic tools for providing privacy and authenticity, respectively, in the public-key setting. Digital signatures are central to the operation of public key infrastructures and many network security schemes (e.g., SSL/TLS, many VPNs, etc.). A public key certificate (also known as a digital certificate or identity certificate) is an electronic document which uses a digital signature to bind together a public key with an identity that is, information such as the name of a person or an organization, their address, and so forth. The certificate can be used to verify that a public key belongs to an individual.

Among the next generation of cryptography solutions there are encryption systems based on an Artificial Neural Network (ANN) with a permanently changing key and non-linear mapping from several input variables to several output variables. The multi-layer topology of the ANN is an important issue, as it depends on the application that the system is designed for. In the future, a trained neural network can replace the above-mentioned techniques for encryption and decryption operations. The benefit of this method includes a higher level of obscurity for the cryptographic process. The drawbacks are unstable implementations, and the vulnerability of the supporting libraries to side-channel attacks.

Modern cryptography technologies, facilitating privacy, have become more accessible and less expensive, and now almost every Internet user worldwide potentially has access to quality cryptography via their browsers (e.g., via Transport Layer Security). The Mozilla Thunderbird and Microsoft Outlook E-mail client programs similarly can transmit and receive emails via TLS, and can send and receive email

---

[23] J. Hea An and T. Rabin, Security for Signcryption: The two-user mode, 2011, last retrieved on 5 December 2018 at https://www.springer.com/cda/content/document/cda_downloaddocument/9783540894094-c1.pdf?SGWID=0-0-45-1017259-p174027366

encrypted with S/MIME, along with special add-ons they use PGP encryption.[24] This raises legal issues with security authorities. While they are protecting information on one hand, they can also be used as a weapon for malicious purposes, such as espionage, transmission of secret criminal information, etc. The investigation of cybercrimes might require a full disclosure of encryption keys to access relevant information.

Global Partners Digital (GPD), which is a social purpose company dedicated to fostering a digital environment underpinned by human rights and democratic values, has developed a global mapping[25] of encryption laws and policies. Thus, the countries with the general write for encryption are Finland, France, Senegal, Malawi, Brazil. Countries with the mandatory minimum or maximum encryption strength are India, Senegal. Among the countries with licensing and registration requirements are Russia, China, Cuba, Morocco, Tunisia, Pakistan, Vietnam, Egypt, Ethiopia, Zambia, South Africa, etc. Countries exercising import and export control are United States, Canada, Belarus, China, Algeria, Tunisia, Morocco, Egypt, Israel, Vietnam, etc. Many countries legally oblige providers in assisting authorities; these are Australia, Russia, China, India, United States, France, Spain, United Kingdom, Finland, Pakistan, Egypt, Nigeria, etc.

Australian Government Passes Contentious Encryption Law[26]

CANBERRA, Australia—The Australian Parliament passed a contentious encryption bill on Thursday requiring technology companies to provide law enforcement and security agencies with access to encrypted communications.

Privacy advocates, technology companies and other businesses had strongly opposed the bill, but Prime Minister Scott Morrison's government said it was needed to thwart criminals and terrorists who use encrypted messaging programs to communicate.

"This ensures that our national security and law enforcement agencies have the modern tools they need, with appropriate authority and oversight, to access the encrypted conversations of those who seek to do us harm," Attorney General Christian Porter said.

Cryptography is also of considerable interest to civil rights supporters, plays a major role in digital rights management and copyright infringement of digital media.

---

[24] A secure searcher for end-to-end encrypted email communication, 2015, http://eprints. maynoothuniversity.ie/7097/1/13013_BALAMARUTHU_MANI.pdf

[25] World map of encryption laws and policies, Global Partners Digital, last retrieved on 13 October 2018 at https://www.gp-digital.org/world-map-of-encryption/

[26] J. Tarabay, Australian Government Passes Contentious Encryption Law, 6 December 2018, last retrieved on 23 December 2018 at https://www.nytimes.com/2018/12/06/world/australia/encryption-bill-nauru.html

2.1.3 Information Access Control

Depending on the security status of the protected information, its access rights are restricted and assigned to the personnel with the corresponding sophistication of the access mechanisms. The same refers to the computer programs, and computer systems that process the information. Authorization of access is based on and is performed through identification and authentication.

Identification verifies the user identity. Before a user can be granted access to the protected network and/or information, it will be necessary to verify that the user is who they claim to be. Typically, this claim is in the form of a username, login ID, personal ID, or a phone number.

Authentication verifies the identity of an individual or system against a presented set of credentials, e.g., the combination of login name and password. The user's login name is the identity presented, and it is verified against a stored form of the password. The future of authentication is a multiple factor method, time based one-time password algorithms, tokens and the extended use of unique biometric identifiers, such as fingerprints, iris scans, and other means based on physical attributes. Such identifiers are reliable, portable, and difficult to clone or fake, given a properly designed authentication system.

Authorization to view, create, change, and delete information is granted after a successful identification and authentication. It is based on the administrative policies and procedures, which allow staff to access specific information and computing services depending on the staff's functions and responsibilities.

There exist several access controls mechanisms. Among them are the Discretionary Access Control—each user controls access to their data, e.g., through Access Control Lists; Mandatory Access Control—a hierarchical approach based on mandatory regulations, where the operating system "decides" who can access the file, i.e., when the security label of the resource and the user's category match; Role Based Access Control, also known as a non-discretionary approach—assigns access as per the staff role within the organization and consolidates access control under the rules of the centralized administration.

Examples of common access control mechanisms in use today include role-based access control, available in many advanced database management systems; simple file permissions provided in the UNIX and Windows operating systems; Group Policy Objects provided in Windows network systems; and Kerberos, RADIUS, TACACS, and the manually configured access lists used in many firewalls and routers.

However, it is worth noting that encryption and access control mechanisms are ineffective and useless if the sensitive information is disclosed by the personnel responsible for these duties.

2.1.4 Social Engineering

Social engineering is one of the most widely spread methods to bypass information security through the direct influence on users. These tricky techniques are employed by cybercriminals to evoke the interest of naive users, consumers or employees, through appealing and intriguing messages, which infect their computers with malware after opening. Their goal is to emotionally interest the target, gain trust through revealing some facts, also confidential information, or to scare or intimidate, and push to pressing the button to open the message and attachment.

Social engineering attacks may be general and specific. General attacks are aimed at providing an entrance to the targeted system or network without any interest in the users. The malware can be delivered through an email, which after being opened by the user, will connect with the attacker. This method is very efficient as it does not require any breach of the permitted defences of the target network. The exploited system can be attacked or used for further reconnaissance and exploitation.

A specific targeted attack is implemented through malware custom-tailored for a specific person or organization. It can be encoded into any attachment, image, video, game, etc. Upon opening and while running these infected files, that is the malware will infect the system.

Social engineering attacks

The cybercriminal will aim to attract the user's attention to the link or infected file— and then get the user to click on it. Examples of this type of attack include:

- The LoveLetter worm that overloaded many companies' email servers in 2000. Victims received an email that invited them to open the attached love letter. When they opened the attached file, the worm copied itself to all of the contacts in the victim's address book. This worm is still regarded as one of the most devastating— in terms of the financial damage that it inflicted.
- The Mydoom email worm—which appeared on the Internet in January 2004—used texts that imitated technical messages issued by the mail server.
- The Swen worm passed itself off as a message that had been sent from Microsoft. It claimed that the attachment was a patch that would remove Windows vulnerabilities. It's hardly surprising that many people took the claim seriously and tried to install the bogus 'patch'—even though it was really a worm.

Source: https://usa.kaspersky.com/resource-center/threats/malware-social-engineering

---

Social Engineering is the most typical and wide-spread threat to organizations and individuals and vigilance is the best protection against it. Raising the awareness of staff can help to draw their attention to suspicious links and always protect their access credentials.

For defence see 2.4 (Defence from selected cyber threats).

2.1.5 Counter Intelligence and Defence Systems

All warfare is based on deception.
Hence, when we are able to attack, we must seem unable;
when using our forces, we must appear inactive;
when we are near, we must make the enemy believe we are far away;
when far away, we must make him believe we are near.

Sun Tzu, 'The Art of War', 500 B.C. [27]

To protect the nation, it is essential that governments and organizations develop and use an effective cyber counterintelligence (CCI) strategy.

> Defending the increasingly complex networks and technology that house and process our sensitive information against sophisticated 21st century threats requires a seamless and well-coordinated four-pronged defensive approach comprising Cyber intelligence (CI), security, information assurance (IA), and cybersecurity professionals working together as a team. Essential is leveraging security and CI disciplines to create mission synergies and extending these synergies into the realm of cyberspace. CI, security, and IA combine in a multidisciplinary approach to provide a more stable network defence posture.
> *Source: The National Counterintelligence Strategy of the United States of America 2016*[28]

The 2016 National Counterintelligence Strategy of the United States of America defines counterintelligence as "information gathered, and activities conducted to identify, deceive, exploit, disrupt, or protect against espionage, other intelligence activities, sabotage, or assassinations conducted for or on behalf of foreign powers, organizations, or persons, or their agents, or international terrorist organizations or activities."[29]

Cyber based espionage activities create a number of key security challenges for security forces and CCI teams to counter. Three of these being (1) the ability for hackers (state-sponsored or otherwise) to access protected information from remote locations; (2) difficultly being able to attribute the attack or data breach to an identifiable perpetrator; and (3) distinguishing cyber exploitation (espionage) from that of covert action (cyber-attacks or acts of force).[30]

---

[27] Sun Tzu, The Art of War, 2016, Canterbury Classics,, Book I, Printers Row Publishing Group (PRPG), San Diego, USA p.4

[28] The National Counterintelligence Strategy of the United States of America 2016, last retrieved on 26 July 2018 at https://www.dni.gov/files/NCSC/documents/Regulations/National_CI_Strategy_2016.pdf

[29] Ibid

[30] D.R. Williams, '(Spy) Game Change: Cyber Networks, Intelligence Collection and Covert Action', George Washington Law Review, June 2011, Vol 79: 4, pp. 1162-1200.

To correctly utilize CCI governments and organizations should understand the underlying motives and objectives of the foreign adversary. These generally entail objectives to (a) penetrate, collect and compromise national security secrets or information to benefit themselves; (b) manipulate and twist facts that are represented of themselves; (c) identify and disrupt national security operations or activities.[31]

CCI includes measures to identify, penetrate, or neutralize computer operations that use cyber weapons as means and mechanism to collect information, focusing on the intrusion and on the intent of the intrusion and tradecraft used.[32] These actions are generally performed through the strategic use of three functional areas of CCI—collection, defence and offense.

Collection of information can be passive, hybrid and active[33].

- Passive—data collected on networks or information systems you have responsibility over. An example would be analysts capturing internal network traffic, collecting system logs, monitoring internal company forums.
- Hybrid—data shared from other networks or information systems or collected from networks designed to provoke adversaries (for example, honeypot data, data sharing between external networks)
- Active—data obtained from external networks or information systems under the influence of an adversary (for example, adversary account or authentication information, interaction on adversary websites). Active data collection usually requires analysts to have access to sensitive data. This type of data collection must be done carefully so that the legal and privacy rights of members are protected.

Each nation creates, monitors, and supports its own cybercrime entities protecting the access to the cyberspace internally and externally. The commercially available cyber tools are used for intelligence gathering, vulnerability assessment, investigations and forensics, internal countermeasures for the illegal use of information technology. However, for high level critical intelligence gathering or defence, specific and enhanced technologies are used as well as high level cyber experts.

---

[31] The National Counterintelligence Strategy of the United States of America 2016, last retrieved on 26 July 2018 at https://www.dni.gov/files/NCSC/documents/Regulations/National_CI_Strategy_2016.pdf

[32] G. Vardangalos, Cyber-Intelligence and Cyber Counterintelligence (CCI): General definitions and principles, Center for International Strategic Analyses', July 2016, last retrieved on 6 June 2018 at https://kedisa.gr/wp-content/uploads/2016/07/Cyber-intelligence-and-Cyber-Counterintelligence-CCI-General-definitions-and-principles-2.pdf

[33] Ibid

As one of the offensive measures of counter intelligence, "honey pots" can be developed and used. These are decoy virtual systems for trapping and deceiving an interested adversary and stimulate malicious activity through "attractive" files and data, while embedded security tools record and report observed activities.

The nations with enhanced cyber capabilities use custom operating systems for advanced protection of strategic data. State agencies in China were using custom Red Flag Linux (up until 2013) as an operating system of internal communications inside ministries with enhanced data protection capabilities. The Turkish government uses Pardus OS (Linux distribution) developed in 2003 by the Turkish Scientific and Technological Research Council. The Russian Army uses Astra (Linux distribution) for Top Secret data.[34] It has been officially certified by the Russian Defence Ministry, Federal Service for Technical and Export Control, and Federal Security Service. The Special Edition version is used in many Russian state-related organizations. Particularly, it is used in the Russian National Centre for Defence Control.

However as good and advanced as all of them are, each of these systems can be vulnerable if basic principles of cyber security are not applied and respected, and the level of enforcement is not strong enough.

2.2 Cyber Security as a Science

At the present state of evolution, an abstract term "information" is merging with Information Systems, which is a socio-technical system, responsible for delivering the information and communication services required by an organization in order to achieve business objectives. An Information System encompasses four components: information (data), people, business processes (procedures), and ICT, which includes hardware, software, and networks.[35]

Cyber security is the field of security that protects information systems from theft or damage to the hardware, software, and to the information on them, as well as from disruption or misdirection of the services they provide. It includes protection against unauthorized network access, data and code injections, intentional or accidental malpractice by operators.

Cyber security, based on computer science was born with the creation of computers and cyberspace, is at the same time a multidisciplinary area

---

[34] Astralinux, last retrieved on 26 July 2018 at http://www.astra-linux.com/products/alse. html

[35] Wilson, Clay, 2007, Information operations, electronic warfare, and cyberwar capabilities and related policy issues. [Washington, D.C.]: Congressional Research Service Report, last retrieved on 16 October 2018 at https://www.history.navy.mil/research/library/online-reading-room/title-list-alphabetically/i/information-operations-electronic-warfare-and-cyberwar.html

which involves mathematics, physics, and electronic engineering, but also law, economics, criminology, philosophy, ethics, psychology, and management. It is aimed at the comprehensive protection of cyberspace, its infrastructures, and covers issues related to its functionality in our life.

The first issue to be considered in the security of cyberspace is the presence of cyber vulnerabilities and their protection.

2.3 Vulnerabilities and Exploits

Vulnerability is a cyber-security term that refers to a flaw in a system that can leave it open to attack. Vulnerability may also refer to any type of weakness in a computer system itself, in a set of procedures, or in anything that leaves information security exposed to a threat.[36]

A computer exploit, or simply an exploit, is an attack on a computer system, especially one that takes advantage of a particular vulnerability the system offers to intruders.[37] Exploit is typically represented as a script of a short tool, which performs a specific set of actions to take advantage of a system flaw and provide additional functionality to the attacker (typically a command line a.k.a. shell).

2.3.1 Vulnerability Overview

Vulnerabilities compromise the confidentiality, integrity or availability of the asset belonging to an individual, organization, or any other involved parties. The consequences depend on the value of the asset and its level of danger to the environment.

Vulnerabilities can be identified through the whole cyberspace related environment and activities, and they can refer to the following areas:[38]

- hardware (susceptibility to humidity, dust, soiling, unprotected storage)
- software (insufficient testing, lack of audit trail, design flaw)
- network (unprotected communication lines, insecure network architecture)
- personnel (inadequate recruiting process, inadequate security awareness)
- environmental (natural disasters, manmade disasters, unreliable power source)

---

[36] Definition—Vulnerability, Techopedia, last retrieved on 26 July 2018 at https://www.techopedia.com/definition/13484/vulnerability

[37] Definition—Exploit, Techtarget, last retrieved on 26 July 2018 at https://searchsecurity.techtarget.com/definition/exploit

[38] ISO/IEC 27005:2018, "Information technology -- Security techniques - Information security risk management", last retrieved on 20 October 2018 at https://www.iso.org/standard/75281.html

- organizational (lack of regular audits, lack of continuity plans, lack of security compliance)

Each of the above categories requires unique approaches in vulnerability management.

The attacker's task is to identify a vulnerability and attack, and usually a single vulnerability is enough. The defender's task is more complicated, and professional vulnerability management is to ensure that all vulnerabilities be proactively identified, classified, remediated and mitigated.

Vulnerabilities have been found in every major operating system including Windows, Mac OS, various forms of Unix and Linux, OpenVMS, and others. The only way to reduce the chance of a vulnerability being used against a system is through constant vigilance, including careful system maintenance (e.g., applying software patches), best practices in deployment (e.g., the use of firewalls and access controls) and auditing (both during development and throughout the deployment lifecycle).[39]

The operating systems are normally designed and scrutinized in testing before release; however, at the current stage flaws are inevitable due to errors in basic security components, such as password verification, account recovery, event history,[40] general system complexity, lack of regular enhancement and upgrades, in certain cases open and accessible connections, elevated privileges, unfiltered ports, protocols, and services.

Common types of software flaws that lead to vulnerabilities include memory safety violations (e.g., Buffer overflows), poor input validation (e.g., Code injection, Directory traversal), privilege confusion (e.g., Clickjacking, Cross-site request forgery in web applications), race conditions (e.g., Symlink races, Time-of-check-to-time-of-use bugs).

Software, if not properly secured, allows attackers to misuse it as an entry point into the system. Programs that do not check user input can allow unintended direct execution of commands or SQL statements (known as buffer overflows, command injection, code injection, or other non-validated inputs).

Some internet websites may contain harmful spyware or adware that can be installed automatically on the computer systems (so called drive-by downloads) when opened.

---

[39] Edited by J.R. Vacca, Computer and Information Security Handbook, 3rd edition, 2017, Morgan Kaufmann Publishers, Massachusetts, USA.

[40] M. Faisal Nagyi, Reinspecting Password, Account Lockout and Audit Policies, 2014, last retrieved on 20 October 2018 at https://www.isaca.org/Journal/archives/2014/Volume-2/Pages/JOnline-Reinspecting-Password-Account-Lockout-and-Audit-Policies.aspx

10 Most Vulnerable OS of the Year 2017[41]

- 10th rank in the list is secured by Microsoft's **Windows 8.1**. The desktop OS was reported for 224 distinguished vulnerabilities in the year 2017. Windows 8.1 ranked 17th in 2016's list of most vulnerable OS with a slightly lesser count of vulnerabilities. In the year 2015, a total number of 154 vulnerabilities were reported for the OS.
- **Windows 7**—another Windows OS and one of the most appreciated desktop OS was reported for 228 vulnerabilities that resulted in securing 9th place in the not so **"good list"**. In the year 2016 it was identified with only 134 vulnerabilities and was in 13th place.
- 8th rank in the list was obtained by Microsoft's **Windows Server 2012**. The server OS was identified with 234 vulnerabilities in the year 2017. The operating system released in September 2012 has been in the top 10 most vulnerable OS lists for the last three years. It ranked 7th and 10th in the list of 2015 and 2016 with most vulnerable OS with almost the same count of reported vulnerabilities (155 and 156 respectively).
- Officially released at Microsoft's Ignite Conference on **September 26, 2016,** the Server OS was reported for 250 different vulnerabilities in the year 2017. The OS ranked 6th in the list of most vulnerable OS of 2017.
- The operating system running on more than 600 million devices ranked fifth on the list. Windows 7 was reported for 266 distinguishable vulnerabilities in the year 2017. The desktop operating system released on July 29, 2015, was at the 9th position last year for 172 vulnerabilities.
- The 4th place in the list goes to Apple's **Mac OS X** for its 299 vulnerabilities that got identified in 2017. Mac OS X, in the year of 2015 was 'topper' of the most vulnerable OS list with its 444 identified vulnerabilities.
- 3rd rank in the list of most vulnerable OS of the year 2017 is obtained by Apple's **iPhone OS**. The OS was reported for 287 distinct vulnerabilities. In the year 2016, Apple's iPhone OS ranked 9th on the list with mere 161 vulnerabilities.
- With more than double the count of vulnerabilities from 2016, Linux Kernel ranked second on the list of most vulnerable OS of 2017. In the year 2017, Linux kernel was reported for 435 vulnerabilities as compared to 217 in 2016. Whereas, in 2015, just 86 vulnerabilities were reported.
- With a total number of 841 identified vulnerabilities in 2017, Google's **Android OS** tops the rank of Operating System with the most number of vulnerabilities. Android got this **"not so desired rank"** second time in a row. Last time as well Android topped this list with its 523 vulnerabilities. In the year 2015, this undesired position was held by Apple's Mac Os X with 444 vulnerabilities.

As an internet connection has a global coverage, and the majority of the users use the same operating systems and applications (Microsoft Windows or MacOS, Microsoft Office, etc.), vulnerabilities require universal identification, classification, and storage, to successfully exchange, study, and develop patches and guidelines for secure programming practices.

---

[41]   10 Most Vulnerable OS Of The Year 2017, 2 January 2018, last retrieved on 16 October 2018 at http://www.cybrnow.com/10-most-vulnerable-os-of-2017/

In 2015, Microsoft called for a better coordinated vulnerability disclosure,[42] stressing that this method pushes software vendors to fix vulnerabilities more quickly and makes customers develop and take actions to protect themselves. Its Coordinated Vulnerability Disclosure (CVD)[43] practice for handling vulnerabilities and Microsoft Vulnerability Research (MSVR) project are responsible for the discovery, reporting, and coordination of vulnerabilities.

The Open Web Application Security Project (OWASP) was established as a not-for-profit charitable organization in the United States on April 21, 2004. It is now an international organization, focused on improving the security of software. It collects a list of vulnerabilities and issues software tools and knowledge-based documentation on application security. Their latest tool projects include OWASP Zed Attack Proxy (ZAP), which automatically finds security vulnerabilities in web applications, OWASP Web Testing Environment—a collection of security tools, OWASP OWTF—security testing tool to find and verify vulnerabilities in short timeframes, etc.[44]

The Mitre Corporation has developed a system called Common Vulnerabilities and Exposures[45], assigning Common Vulnerability Exposure ID (CVE ID) to vulnerabilities affecting products within their distinct, agreed-upon scope, for inclusion in first-time public announcements of new vulnerabilities. These CVE IDs are provided to researchers, vulnerability disclosers, and information technology vendors. CVE Numbering Authorities (CNAs) are organizations from around the world that are authorized to assign CVE IDs.

> The 2018 Mid Year VulnDB QuickView report of the Risk Based Security shows there have been 10,644 vulnerabilities disclosed through June 30th. This is the highest number of disclosed vulnerabilities at the mid-year point on record. The 10,644 vulnerabilities catalogued during the first half of 2018 by Risk Based Security's research team eclipsed the total covered by the CVE and National Vulnerability Database (NVD) by well over 3,000.[46]

---

[42] Betz Ch., A Call for Better Coordinated Vulnerability Disclosure, 11 January 2015, last retrieved on 15 August 2018 at https://blogs.technet.microsoft.com/msrc/2015/01/11/a-call-for-better-coordinated-vulnerability-disclosure/

[43] Coordinated Vulnerability Disclosure, MSRC, last retrieved on 21 October 2018 at https://www.microsoft.com/en-us/msrc/cvd

[44] The OWASP Foundation, last retrieved on 13 October 2018 at https://www.owasp.org/index.php/Main_Page

[45] CVE Numbering Authorities, last retrieved on 15 August 2018 at https://cve.mitre.org/cve/cna.html

[46] More Than 10,000 Vulnerabilities Disclosed So Far In 2018—Over 3,000 You May Not Know About, 13 August 2018, last retrieved on 17 August 2018 at https://www.riskbasedsecurity.com/2018/08/more-than-10000-vulnerabilities-disclosed-so-far-in-2018-over-3000-you-may-not-know-about/

The vulnerability disclosure can be made by the manufacture (self-disclosure), or disclosed by a third party, who informs the manufactures, or vendors.

For the safety and security of the public and the vendor reputation, responsible disclosure best practices recommend that the vulnerability finder first notify the vendor privately. The finder must provide the vendor with adequate time not only to develop fixes and/or patches but also to push out software updates. Once fixes, patches, and updates have been proliferated throughout the software supply chain, only then should the finder publicly disclose the findings.[47]

Google Discloses Windows Zero-Day Before Microsoft Can Issue Patch[48]
There's a tiny scandal brewing among two of Silicon Valley's elites after Google engineers have publicly disclosed a zero-day vulnerability affecting several Windows operating system versions before Microsoft could issue a patch to address the issue.

At the heart of the problem is a series of exploitation attempts detected by Google's Threat Analysis Group, a division of Google's security team that keeps an eye out for complex and out-of-the-ordinary cyber-attacks, usually specific to state-sponsored cyber-espionage groups.

On October 21, Google's engineers said they've detected a sophisticated attack routine that combined two zero-days, one in Adobe Flash Player and the other affecting all Windows versions between Vista and Windows 10.

Currently, there is no definite agreement on the reasonable amount of time to allow a vendor to patch vulnerabilities before full public disclosure. Thus, CERT/CC commits to disclose vulnerabilities to the public 45 days after the initial report to CERT, regardless of the existence or availability of patches or workarounds from affected vendors. Extenuating circumstances, such as active exploitation, threats of an especially serious (or trivial) nature, or situations that require changes to an established standard may result in earlier or later disclosure. Disclosures made by the CERT/CC will include credit to the reporter unless otherwise requested by the reporter. CERT will apprise any affected vendors of their publication plans and negotiate alternate publication schedules with the affected vendors when required.[49]

---

[47] Navigating responsible vulnerability disclosure best practices, 1 December 2017, last retrieved on 19 October 2018 at https://www.synopsys.com/blogs/software-security/responsible-vulnerability-disclosure-best-practices/

[48] Catalin Cimpanu, Google Discloses Windows Zero-Day Before Microsoft Can Issue Patch, 2 November 2016, last retrieved on 17 October 2018 at https://www.bleepingcomputer.com/news/security/google-discloses-windows-zero-day-before-microsoft-can-issue-patch/

[49] Vulnerability Disclosure Policy, 19 February 2018, last retrieved on 12 October 2018 at https://vuls.cert.org/confluence/pages/viewpage.action?pageId=30638083

> **Companies v. US Government and the Public Disclosure Debate**[50]
>
> In 2017 and 2018, there was another high profile case of questionable vulnerability disclosure practices known as the Spectre and Meltdown chip flaws. In January of 2018, Intel revealed that millions of their computer chips were vulnerable to hacking; however, Intel did not go public with this information when they discovered it in June of 2017. Instead, Intel told select vendors about the problem (Huawei, Google, Alibaba, and Lenovo, etc.) while they worked behind the scenes to fix it.
>
> In this case, certain companies were working together to address the problem, but in July of 2018, several U.S. Senators pointed out that companies with close Chinese government ties knew about the vulnerability before the U.S. government did, putting national security and consumers' security at risk.

Most industry vendors generally agree that a 90-day deadline is acceptable. In 2010, Google recommended a 60-day deadline to fix a vulnerability before full public disclosure, seven days for critical security vulnerabilities, and fewer than seven days for critical vulnerabilities that are being actively exploited. However, in 2015 Google extended that deadline to 90 days for its Project Zero program.[51]

2.3.2 Exploits

An exploit takes advantage of a weakness in an operating system, application or any other software code, including application plug-ins or software libraries.[52]

Exploits can be classified based on the following:

1. The mode of work:
   - A remote exploit: exploits the security vulnerability without any prior access to the system
   - A local exploit: requires a prior access to the system and is used to increase the privileges
2. Type of attacks they are able to accomplish:
   - unauthorized data access
   - arbitrary/remote code execution
   - denial of service
   - privilege escalation
   - malware delivery

---

[50] The Vulnerability Disclosure Debate, 20 August 2018, Markkula Center for Applied Ethics, last retrieved on 12 October 2018 at https://www.scu.edu/ethics/focus-areas/business-ethics/resources/the-vulnerability-disclosure-debate/

[51] Feedback and data-driven updates to Google's disclosure policy, 13 February 2015, last retrieved on 17 July 2018 at https://googleprojectzero.blogspot.com/2015/02/feedback-and-data-driven-updates-to.html

[52] Exploit—definition, Techtarget, last retrieved on 17 October 2018 at https://searchsecurity.techtarget.com/definition/exploit

3. Type of vulnerability being exploited:
   - buffer overflow exploits
   - code injection
   - side-channel attacks

Exploits can provide a step by step access to the system. In this case they start from low level and escalate the privileges, if the vulnerability is found in a low-privileged third-party application. They can also provide a maximum possible access to a computer system through administrative level access, if the vulnerability is found in a critical component, such as operating system kernel or file-sharing protocol.

One year later: EternalBlue exploit more popular now than during WannaCryptor outbreak[53]

It's been a year since the WannaCryptor.D ransomware (aka WannaCry and WCrypt) caused one of the largest cyber-disruptions the world has ever seen. And while the threat itself is no longer wreaking havoc around the world, the exploit that enabled the outbreak, known as EternalBlue, is still threatening unpatched and unprotected systems. And as ESET's telemetry data shows, its popularity has been growing over the past few months and a recent spike even surpassed the greatest peaks from 2017.

The EternalBlue exploit targets a vulnerability (addressed in Microsoft Security Bulletin MS17-010) in an obsolete version of Microsoft's implementation of the Server Message Block (SMB) protocol, via port 445. In an attack, black hats scan the internet for exposed SMB ports, and if found, launch the exploit code. If it is vulnerable, the attacker will then run a payload of the attacker's choice on the target. This was the mechanism behind the effective distribution of WannaCryptor.D ransomware across networks.

EternalBlue has enabled many high-profile cyberattacks. Apart from WannaCryptor, it also powered the destructive Diskcoder.C (aka Petya, NotPetya and ExPetya) attack in June 2017 as well as the BadRabbit ransomware campaign in Q4 2017. It was also used by the Sednit (aka APT28, Fancy Bear and Sofacy) cyber-espionage group to attack Wi-Fi networks in European hotels.

Zero-day exploits aim at zero-day vulnerabilities which contain a critical security vulnerability yet unknown to the vendor and public. After an exploit is disclosed, the vulnerability is fixed through a patch and the exploit becomes unusable against all the applications of the patched version or higher.

Zero-day exploits can also be supplied as a part of an existing exploit kit, a collection of exploits, tools, and a framework (interface) to manage them. Those toolkits contain a variety of tools ranging from vulnerability scanners to automated exploitation of a range of targets at once. Exploit kits act as a repository and make it easy for users without any technical

---

[53]   Ondrey Kubovic, One year later: EternalBlue exploit more popular now than during WannaCryptor outbreak, 10 May 2018, last retrieved on 15 August 2018 at https://www.welivesecurity.com/2018/05/10/one-year-later-eternalblue-exploit-wannacryptor/

knowledge to launch cyberattacks. Advanced users can add their own exploits to toolkits and use them alongside the pre-installed ones.

Automated exploits are often composed of two main components: the exploit code and the shell code. The exploit code is the software that attempts to exploit a known vulnerability. The shell code is the payload of the exploit—software designed to run once the target system has been breached.

As an example, an exploit can be initiated from a web-site visited by the user accidently or through a link in the phishing email. This web-site can have an exploit kit with malicious software that can launch an attack on the victim's computer. A well automated exploit kit gathers information on the victim's machine, finds vulnerabilities, determines the appropriate exploit, delivers the exploit, and further runs a post-exploitation module to maintain further remote access to the compromised system. To cover up tracks, it uses special techniques like erasing logs to avoid detection. Kits may have a Web interface showing active victims and statistics. They may have a support period and updates like commercial software.

Current most well-known exploit kits are MPack, Phoenix, Blackhole, Crimepack, RIG, Angler, Nuclear, Neutrino, and Magnitude.[54]

2.4 Defence from Selected Cyber Threats

The best defence is a good offence

A good defence ensures the resilience of the cyber system to external and internal threats, and is based on comprehensive security solutions, regular cyber hygiene, simulation exercises and after-action reviews. It designs efficient cyber architecture, identifies vulnerabilities, prevents exploits, timely detects and stops attacks, applies coercive strategies when needed, etc.

2.4.1 General Defence Solutions

Following the doctrine "to prevent is always better than to cure", organizations develop cyber security strategies, policies and procedures to ensure prevention of any malicious activity, information leaks or damage to internal systems.

The following attack vectors should be considered in developing defence strategies:

1. Cyber-attack: the attackers have no physical access to systems or devices;

---

[54] Exploit Kit—Definition, Trent Micro, last retrieved on 19 October 2018 at https://www.trendmicro.com/vinfo/sg/security/definition/exploit-kit

2. Physical attack: the attackers have no digital access to systems or devices;
3. Cyber-enabled physical attack: security system is digitally compromised to enable physical access for the attackers;
4. Physical-enabled cyber-attack: physical actions allow remote access to the previously unreachable computer systems or networks (e.g., rogue device is physically delivered and connected to the internal network).

In cyber-attacks per se, there is no physical contact between the attacker and the target, e.g., BlackEnergy3, NotPetya, etc.

In physical attacks the attackers physically access the facility to interact with the system. The physical access to the hardware, communication wire layout and the internal wireless network of the facility may give attackers the level of access to the network sufficient to conduct a successful attack without the exploitation of corporate software, e.g., access an unlocked computer, acquire access credentials from employees or their notes, etc.

In cyber-enabled physical attacks, the security system is compromised to enable easy physical access for attackers. The security systems may be disabled remotely for the physical attack to be conducted, e.g., security cameras disabled to allow undetected or uninterrupted physical access to the facility, RFID cloning allows them to copy electronic entrance cards, etc.

In physical-enabled cyber-attacks, physical actions allow access to the computer system or network, e.g., a wired network physical access (wiretapping), wireless network physical access (long range antennas, WiFi attack vectors, BYOD, etc.), an insider threat which can impose limitations on the best cyber security solutions.

One of the most effective defence modes is the defence in depth system with a multilayer security control placed throughout an ICT system. It addresses security vulnerabilities in personnel, technology, and operations for the duration of the system's life cycle.[55] Additional protection is provided in case one of the previous layers is breached. This mitigates the risks of vulnerabilities through human errors, technologies, operations and provides more time to respond to attacks.

Vulnerability assessment and penetration testing allows the owners of the system to discover the weaknesses in systems and networks. Penetration testing methodologies have different recommendations and approaches to the simulated cyber offensive operations: e.g., OWASP, OSSTMM, NIST 800-115, ISSAF, "KillChain", etc. They can be performed

---

[55] Edited by J.R. Vacca, Computer and Information Security Handbook, 3rd edition, 2017, Morgan Kaufmann Publishers, Massachusetts, USA.

from a "white box" perspective, in which the information about the "target" environment is provided to the contractors or can be done from a "black box" perspective, in which no information about the target network is revealed. "White box" testing uses fewer resources and is less time-consuming, while "black box" testing mimics a real-life external attack scenario, and thus requires more time and resources.

The basic software solutions that assist in providing cyber security measures for corporate and industrial networks include antivirus software, firewalls, intrusion detection and protection systems, and honeypot implementations.

Anti-virus programs are not always effective against new viruses, especially in the CBRNe environment, where low-power devices do not allow additional software (such as antiviruses) to be easily deployed. The malware developers test their new malware using the major antivirus applications available and online services, some of which are available for free. For example, new strains of malware are often discovered by online malware analysis websites (such as VirusTotal) when the malware developers test it for detectability.

Firewalls defend the network perimeter from external attacks. Additional firewalls are deployed between networks to provide additional security and to make it harder for the attackers to move from one network to another.

Intrusion detection and prevention systems (IDS) are primarily focused on identifying possible incidents in systems and networks, logging information, reporting attempts and learning. IDS have become a necessary addition to the security infrastructure of nearly every organization.

Honeypots are used to redirect the attacker's attention away from critical devices in the network. Multiple honeypots can be deployed to create a virtual illusion of the industrial network that is not connected to any devices. Manipulations with such networks do not affect the facility in any way, while keeping the attackers busy with tracking false paths.

Currently firewalls, antivirus software and IDS have become the norm for the organizations. In highly secure environments these may not be sufficient. The same is true in relation to sophisticated attack vectors.

Below are the examples of defence against the most wide-spread attack techniques:

- Phishing
- Direct Hacking
- Bots and Botnets
- Password Guessing
- Ransomware

2.4.2 Cyber Defence from Phishing

Phishing emerged more than two decades ago and has become a well-developed type of cyberattack. Authentic communication and fake messages are getting harder to distinguish, especially with AI-enhanced data mining and email generation tools. Human factors are the main reason such attacks succeed.

Due to technological progress and growing interconnectivity the perfection of phishing techniques leads to the growing list of victims, that is users who over rely on antivirus systems, ignore the warning signals, get inspired by exciting messages and click on attachments or links out of curiosity. In 2015, there were almost 100,000 reports of Phishing emails being received in the UK, and it is believed that 50% of these attempts were successful.[56]

> In 2017, Kaspersky Lab's anti-phishing system was triggered more than 246 million times, representing a 59% increase compared to 2016. These included fake lottery wins that supposedly pay out in bitcoins, but instead of providing details to collect the prize, the users were actually passing their details on to scammers.
> Another campaign involved emails pretending to offer bitcoin mining tools or trading instructions. The emails contained attachments that masqueraded as the advertised service but were no more than shells to deliver malware.[57]

Phishing attacks may be addressed to millions, inspiring users to open an attachment or go to a malicious web page. They can be also custom-tailored and addressed to a specific person (spear phishing).

> **New tool automates phishing attacks that bypass two-factor authentication[58]**
> A new penetration testing tool published at the start of the year by a security researcher can automate phishing attacks with an ease never seen before and can even blow through login operations for accounts protected by two-factor authentication (2FA). From a report:
> Named Modlishka—the English pronunciation of the Polish word for mantis -- this new tool was created by Polish researcher Piotr Duszynski. Modlishka is what IT professionals call a reverse proxy, but modified for handling traffic meant for login pages and phishing operations. It sits between a user and a target website -- like Gmail, Yahoo, or ProtonMail. Phishing victims connect to the Modlishka server (hosting a phishing domain), and the reverse proxy component behind it makes requests to the

---

[56] Top 5 biggest phishing scams, 3 June 2016, last retrieved on 21 July 2018 at https://www.theinquirer.net/inquirer/feature/2460065/top-5-biggest-phishing-scams
[57] Phishing attacks increased by 59% in 2017, 2 March 2018, last retrieved on 15 August 2018 at https://www.itgovernance.eu/blog/en/phishing-attacks-increased-by-59-in-2017
[58] New Tool Automates Phishing Attacks That Bypass 2FA, 9 January 2019, last retrieved on 10 January 2019 at https://it.slashdot.org/story/19/01/09/1817234/new-tool-automates-phishing-attacks-that-bypass-2fa?utm_source=rss1.0mainlinkanon&utm_medium=feed

> site it wants to impersonate. The victim receives authentic content from the legitimate site --let's say for example Google—but all traffic and all the victim's interactions with the legitimate site passes through and is recorded on the Modlishka server.

The most commonly recommended preventative measure is to avoid opening emails or links that look suspicious, unauthentic, and ambiguous. Currently, there is a growing market for alternative AI-based anti-spam and anti-phishing solutions to mitigate this threat by technical means.

2.4.3 Cyber Defence from Exploitation

Exploitation always involves specific software and their discovered vulnerability. Secure programming practices advises developers to avoid vulnerable input methods (certain operators in programming languages), and employ software hardening and input sanitization. Without these measures attackers use available input methods to inject malicious code into the software and cause the program to perform unintended actions (e.g., halt or provide unauthorized access to the system). To defend against such exploitation, should a vulnerability be found in the software, the software developers have to issue and distribute timely patches for the discovered vulnerabilities and users have to deploy them accordingly.

A patch is a set of modifications to a computer program or its supporting data designed to update, fix, or improve it. This includes mitigating security vulnerabilities and errors, and improving the usability and performance. Poorly designed patches can sometimes cause new vulnerabilities and version incompatibility.

The current common practice is to use the available software tools in the discovery of vulnerabilities in a computer system, which provide automate routine checks and version analysis. For instance, vulnerability scanners as Nessus, GFI LANguard, Retina, Core Impact, X-scan, Internet Security Systems Internet Scanner, etc.[59]

2.4.4 Cyber Defence from Bots and Botnets

A Bot is an efficient receiver and transmitter malicious program designed to scan a system, find specific information such as credit card information, weak points in new software patches or previously unknown access points that can then be exploited. Large-scale networks of such bots are called botnets. Computers can be infected by bots, and can be targeted by botnets.

Most of the bots in the persistence mode of operation (start every time when the operating system starts) keep an open port and check a certain

---

[59] Top 125 Network Security Tools, last retrieved on 17 October 2018 at https://sectools. org/tag/sploits/

website (hosted by the attackers) if there are any new directives. This type of bots is used for spam and DDoS attacks.

The primary solution to remove bots from the system is to deploy an anti-malware solution. The following solutions against being a target for botnets are used by the organizations:

- Anti-spam solutions allow for the reduction of the volume of advertisements and fraudulent emails, received by companies.
- Anti-DDoS solutions block suspicious traffic, which threaten to overload the external network connections.
- Load-balancing solutions for data centres allow for the reduction of the volume of traffic on a single node in the network, as well as preventing bruteforce (password guessing) attacks.

Defence against concerted attacks, such as large-scale Distributed Denial of Service (DDoS) attacks, at this point usually revolve around waiting till the attack exhausts itself, either by bringing redundant systems online, or some similar strategy. Web application firewalls and anti-DDoS services (e.g. Cloudflare), created to defend websites against DDoS attacks, allow for the reduction of the risk of overload.

2.4.5 BYOD (Bring Your Own Device)

A study conducted by HP Inc. in 2014 identified that 97% of employees' devices contained privacy issues and 75% did not have sufficient data encryption.[60] This refers to storage devices, laptops, smartphones, tablets, and wearable technology. Many organizations have created BYOD policies to allow employees to use personally owned devices and as well as to mitigate the risks they present to the internal network. If a personal device is compromised by a third party then the company's software could be further compromised, allowing attackers access to sensitive data.

The most common way of addressing those risks is to ensure the BYOD policy contains additional security measures, installed on any devices. This increases protection measures for the personal device, as well as increases the security of the organizational internal network.

2.4.6 Password Guessing

If the exploitation phase fails, the attackers will repeatedly attempt different password combinations, in order to guess a legitimate login and password combination, that could be used to enter the system without exploitation.

---

[60] HP Extends Persistent Data Protection from Data Center to Cloud with Encryption Innovations, 10 June 2014, last retrieved on 21 October 2018 at http://www8.hp.com/us/en/hp-news/press-release.html?id=1695191

Generally accepted guidance for creating a strong password includes:

- Avoid using personal information (names, cities, or date of birth)
- Ensure all passwords are significantly different by adding unusual symbols
- Avoid using the same password for different applications
- Use strong passwords (over 16 symbols)[61]

The alternative would be to use biometric authentication, token-based authentication (one-time passwords), or dual-factor authentication.

With a strong password policy ensured in the company, the success of password guessing is minimal.

2.4.7 Cyber Defence from Ransomware

Ransomware is a type of malicious software designed to block access to a computer and its files until a ransom is paid out, it has the ability to cause massive disruption to productivity within the workplace. Previously ransomware was used to block only a specific user account, and it was possible to remove it from a different user profile or by operating the system's "safe mode". Current ransomware encrypts user files first, and holds them for ransom, promising to return the decryption key once the ransom is paid.

Organizations (both private and public) can become a target of a ransomware attack. The following security practices can be recommended to defend against ransomware:

1. Critical files should be frequently backed up. Backup activity should be isolated to mitigate unauthorized access.
2. Received and downloaded archived files should be scanned using anti-malware programs before opening them.
3. Cyber-Hygiene should be exercised to avoid phishing emails and websites.

A March 2018 report[62] by Kaspersky Lab highlighted how one cryptomining gang tracked over six months mined 7 million USD with the help of 10,000 computers infected with mining malware. Meanwhile, the amount of new ransomware variants fell from 124,320 in January to 71,540 in March, decreasing by 42 percent.

While big companies take cybersecurity more seriously, the growing number of malicious applications means that some small businesses go bankrupt due to ransomware attacks. In fact, 20 percent of businesses

---

[61] Password Security: Complexity vs. Length [Updated 2018], 10 September 2018, last retrieved on 25 October 2018 at https://resources.infosecinstitute.com/password-security-complexity-vs-length/

[62] Cryptominers gain ground, 27 June 2018, last retrieved on 25 October 2018 at https://www.kaspersky.com/blog/cryptominers-almost-double/22898/

still don't have a disaster recovery solution. Which means that when a malicious attack comes—and it will—one-fifth of businesses have no method or plan for recovering data, applications, customer information, servers, or systems. And 42 percent of the businesses that do have a disaster recovery strategy use a tape-based, outdated backup method.[63]

In 2018, DRaaS (Disaster Recovery as a Service) solutions are considered to be the most popular solution against ransomware attacks. It provides the deployment of backup files and configurations (typically prepared in advance), identification of which backups were uninfected and not corrupted, and to restore the functionality of the system and network.

At the RSA Conference in 2018,[64] it was reported that ransomware attacks, known to be the most popular type of malware in 2017, declined significantly in volume over the past three months of 2018. Meanwhile, cryptocurrency mining is overtaking the malware market.

2.4.8 Cyber Defense from Cryptomining Malware

Cryptocurrency mining malware uses computer resources for the resource-heavy mathematical computations, required for generation of cryptocurrency to benefit malicious actors.

Cryptominers are similar to any traditional malware; however, they are aiming to use as much system capacity (CPU or GPU) as possible and thus loading system resources to the highest possible levels, creating heavy noise from cooling fans to counter overheating, shutting down resource-heavy user applications, and significantly slowing down any other running processes.

> Crypto-mining malware activity grew significantly in the first quarter of 2018, according to new research, suggesting that threat actors are finding this tactic to be more lucrative than traditional ransomware attacks due to the increasing popularity and value of digital currencies.[65]
>
> According to the McAfee Labs Threats Report, researchers observed more than 2.9 million samples of crypto-mining malware in the first quarter of 2018—a 629 percent increase from just 400,000 samples in the last quarter of 2017.[66]

---

[63] W. Ashford, One in five businesses hit by ransomware are forced to close, 3 August 2016, last retrieved on 24 May 2018 at https://www.computerweekly.com/news/450301845/One-in-five-businesses-hit-by-ransomware-are-forced-to-close-study-shows

[64] Cryptominers Leaped Ahead of Ransomware in Q1 2018. Comodo Cybersecurity Threat Research Labs' Global Malware Report, RSA Conference 2018, 17 April 2018, last retrieved on 18 October 2018 at https://www.comodo.com/news/press_releases/2018/04/cryptominers-leaped-ahead-of-ransomware-comodo-labs-report-q1-2018.html?

[65] Does the Rise of Crypto-Mining Malware Mean the End of Ransomware? 9 July 2018, last retrieved on 27 July 2018 at https://securityintelligence.com/news/does-the-rise-of-crypto-mining-malware-mean-the-end-of-ransomware/

[66] McAfee Labs Threats Report, June 2018, last retrieved on 19 October 2018 at https://www.mcafee.com/enterprise/en-us/assets/reports/rp-quarterly-threats-jun-2018.pdf

Countermeasures include anti-malware solutions and specialized anti-cryptomining addons for browsers.

2.4.9 Recovery and Reporting

Cyberattacks are followed by a response and recovery procedure as per the established security policies, or emergency needs.

As per the routine response plan a team of experts evaluates the damage, and identifies the cause of the breach, reviews the event logs, generates conclusions, and engages the recovery procedures.

In complicated cases, an external support of high level and legal experts can be required to resolve potential non-technical issues (e.g., reputation loss, compromised customer information, legal claims).

Recovery measures include using proactive backups of critical data, load balancing, backup workstations, domain user accounts, etc. Possible responses to a security threat or risk are:

- reduce/mitigate—implement safeguards and countermeasures to eliminate vulnerabilities or block threats
- assign/transfer—place the cost of the threat onto another entity or organization such as purchasing insurance or outsourcing
- accept—evaluate if the cost of the countermeasure outweighs the possible cost of loss due to the threat

After-action reports and lessons learned reinforce security measures and policies, thus mitigating future security risks and reducing the possibility of emergency situations. For the organizations with a well-developed response and recovery plan, and resilient systems, cyberattacks are not critical to the routine operations.

2.4.10 Forensics Investigation

Digital forensics investigates data and traces found in the compromised digital devices during or after a suspected cyberattack. The goal is to establish if the cyberattack took place and attribute it by presenting data evidence from computers and other digital media. The basic elements include collecting, preserving and analysing data evidence, investigating and reporting.

Digital forensics specialists are qualified experts who document legal statements for the court. It is essential that the digital investigation should be conducted by highly skilled data recovery experts with a legal background, and that non-qualified personnel do not damage the evidence. The investigation must be documented, and the conclusion should comply with the legal standards.

> **Digital Forensics "Rule of Thumb"** is applied when the computer is discovered at the crime scene:
> - If the computer is turned on, do not turn it off until the digital forensics experts arrive.
> - If the computer is turned off, do not turn it on.
>
> Criminals use stateless operating systems and software (which remove all history and settings as soon as they are closed) not to leave digital traces of their activities. Turning the computer off will remove information loaded to the Random Access Memory and delete the potential evidence.
>
> Turning the unpowered computer on may result in data destruction (physical or virtual), if the criminals installed any electronic devices or specific full-disk encryption software that deletes all the local data if the password is not correct.

There are certain best practices for digital forensics analysts:

- Create a cryptographic hash digest of the original media (MD5/SHA-1). If anything in the file or on the hard-drive changes the hash changes. This allows copies of files to be used in court as authentic original evidence.
- Using forensics tools (e.g. Encase, Forensic Tool Kit, Helix) to analyse the evidence documenting the findings. These tools can discover data fragments. Open source tools may not be acceptable in court.

Reporting is a mandatory process for any forensics expertise. The report should be comprehensive, include satisfactory conclusions and documented evidence, signed by a qualified digital forensics specialist.

The enhanced defence measures are specifically relevant to critical infrastructure and CBRNe assets.

2.5 CBRNe Cyber Security

CBRNe assets require an enhanced national level protection and any attack against them can be treated as equal to a warfare operation.

Technologically, by 2030, there will be lower obstacles to the covert development of nuclear weapons and to the development of more sophisticated nuclear weapons. Chemical and biological weapons are likely to be 1) more accessible to both state and nonstate actors due to lower barriers towards the acquisition of current and currently emerging CBW technologies; 2) more capable, particularly in terms of their ability to defeat current or currently emerging defensive countermeasures; 3) more discriminate; that is, more precisely targeted and/or more reliably low- or nonlethal; and 4) harder to attribute (utilizing hitherto unknown agents and/or delivery mechanisms) than the traditional forms known today. No major new technological developments regarding radiological weapons are foreseen.[67]

2.5.1 CBRNe Cyber Vulnerabilities

The merging of automation and computer systems with industrial control systems transferred cyber threats to the CBRNe industry. They range from spear phishing and social engineering to custom designed exploits and deliberately planted routers.

Most computerized CBRNe systems are disconnected and isolated from the Internet. However, those measures have proven to be insufficient against high-level attackers. The vulnerabilities associated with control systems relate to their production, maintenance, operations, supply chain, communications, etc. (see Figure 2.1).

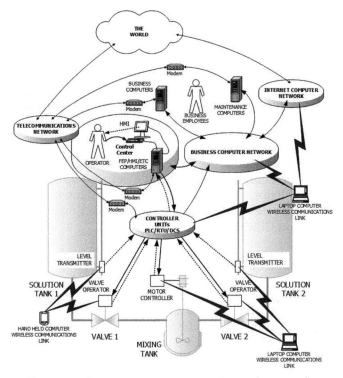

Figure 2.1: Communications access to control systems[68]

---

[67] John P. Caves, Jr. and W. Seth Carus, The Future of Weapons of Mass Destruction and Their Nature and Role in 2030, retrieved on 9 June 2018 at http://ndupress.ndu.edu/ Portals/97/Documents/Publications/Occasional%20Papers/10_Future%20of%20WMD. pdf

[68] Overview of Cyber Vulnerabilities, ICS-CERT, last retrieved on 17 October 2018 at https://ics-cert.us-cert.gov/content/overview-cyber-vulnerabilities#under

The key computerized CBRNe devices include ICS, which have a multi-layered structure and are comprised of Supervisory Control and Data Acquisition (SCADA) systems, Distributed Control System (DCS), Human-Machine Interfaces (HMIs), Master Terminal Units (MTUs), Programmable Logic Controllers (PLCs), Remote Terminal Units (RTUs), Intelligent Electronic Devices (IEDs), and other items.[69] All of them can be vulnerable to cyberattacks.

In a typical large-scale industrial facility many computer, controller and network communications components are integrated to provide for the operational needs of the entire industrial cycle. The typical network architecture is shown in Figure 2.2.

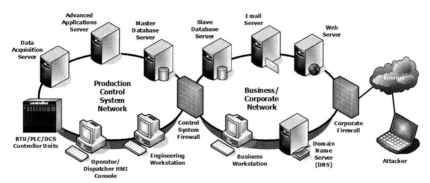

Figure 2.2: Typical two-firewall network architecture[70]

ICS, designed decades ago, were meant to provide uninterrupted services and their architecture is static. Most software used in ICS lack authentication in the design, while ICS equipment relies on physically security. Any possible connection link to the internet can be perceived as a vulnerability. Moreover, information on the functionality of the ICS-specific protocols and proprietary interfaces are publicly available, allowing malicious actors to develop potential attack vectors and methods.

During maintenance, physical access to the CBRNe electronic systems enables attackers' advanced access to the internal network and to potentially allow a direct physical extraction of the data storage devices and equipment (USB drives, laptops, hard drives, etc.).

> In 2016, the number of attacks targeting ICSs increased over 110 percent compared to 2015, and the 2017 SANS study[71] found that 69 percent of ICS security practitioners believe threats to the ICS systems are high or severe and critical.

---

[69] Ibid

[70] Ibid

[71] 2017 Threat Landscape Survey: Users on the Front Line, 2017, last retrieved on 21 October 2018 at https://www.sans.org/reading-room/whitepapers/threats/2017-threat-landscape-survey-users-front-line-37910

The modern CBRNe equipment is designed with the enhanced security measures in mind. However, after a significant timeframe for assessment and verification, the security measures become outdated and further updates are required upon installation. Due to the strategic character and numerous approval levels, identification and patching of the new vulnerabilities in control systems is time consuming. In practice, the updates are rarely deployed by the operators of the equipment, preserving the systems vulnerabilities. The new solutions cannot be easily addressed by introducing a new "application level" software, as both hardware and software architectures do not allow adaptive cyber security activities to duly perform operational security.

More adaptive and dynamic systems are required to meet constantly evolving cyber threats. Until the next generation's protocols are developed and implemented worldwide, enhanced vulnerability management should provide support to the CBRNe security architecture.

As per the forecasts, the CBRNe danger will be only increasing over time. Caves and Carus in "The future of weapons of mass destruction and their nature and role in 2030" predict: "Technologically, by 2030, there will be lower obstacles to the covert development of nuclear weapons and to the development of more sophisticated nuclear weapons.

Attacks on CBRNe facilities can lead to national and even international level damage and irreversible consequences. In this regard, the stockpiled nuclear weapons, if compromised by a high level cyberattack, represent the danger of the existential level.

2.5.2 Vulnerabilities of Nuclear Weapon Systems and N3C

Today's NSA secrets become tomorrow's Ph.D. theses
and the next day's hacker tools

Bruce Schneier

Nuclear Weapons Overview

Nuclear weapons, enhanced by computerized systems, are becoming more dependent on virtual intelligence. Though ensuring the greatest deterrence of all time, with their increased complexity and embedded cyber equipment, they have become vulnerable to stealthy, high speed, unpredictable and anonymous cyberattacks and have expanded the range of attack vectors.

> Nuclear weapon means any weapon that derives its destructive force from nuclear reactions. The nuclear reaction may either be fission or a combination of fission and fusion. Nuclear weapons are considered weapons of mass destruction.[72]

---

[72] Nuclear Weapon Law and Legal Definition, last retrieved on 9 December 2018 at https://definitions.uslegal.com/n/nuclear-weapon/

Cyberattacks on the Nuclear Command, Control, and Communications (NC3) systems could cause false alarms, interrupt critical communications, interfere with the delivery systems, or even lead to full control of a nuclear weapon, targeted launch or detonation within the silo.

It is widely believed that the nuclear systems have the safest design and highest level of protection. However, the human factor remains present and human errors and unauthorized use with the systems, neither can be excluded with errors in design, calculations or equipment failures. Even previously secure air-gapped solutions are no longer the guarantee of protection from cyberattacks.[73]

> There are a number of vulnerabilities and pathways through which a malicious actor may infiltrate a nuclear weapons system without a state's knowledge. Human error, system failures, design vulnerabilities, and susceptibilities within the supply chain all represent common security issues in nuclear weapons systems. Cyberattack methods such as data manipulation, digital jamming and cyber spoofing could jeopardize the integrity of communication, leading to increased uncertainty in decision-making.[74]
>
> *Cybersecurity of Nuclear Weapons Systems Threats,*
> *Vulnerabilities and Consequences,*
> Beyza Unal and Patricia Lewis, Chatham House, 2018

Recognizing the need of the nuclear systems for peaceful purposes, existence and the continuing production of nuclear weapons and their support equipment for military deterrence, the issue of their protection from sophisticated cyber threats is becoming one of paramount importance.

> **Hackers are targeting nuclear facilities, homeland security dept. and F.B.I. say[75]**
> Since May 2017, hackers have been penetrating the computer networks of companies that operate nuclear power stations and other energy facilities, as well as manufacturing plants in the United States and other countries.
> Among the companies targeted was the Wolf Creek Nuclear Operating Corporation, which runs a nuclear power plant near Burlington, Kan., according to security consultants and an urgent joint report issued by the Department of Homeland Security and the Federal Bureau of Investigation last week.

Information about cyberattacks on nuclear systems is very sensitive and mostly classified, and not available openly. However, it might be

---

[73] S. Abaimov and P. Ingram, Hacking UK Trident: The Growing Threat, British-American Security Information Council 2017

[74] Cybersecurity of Nuclear Weapons Systems Threats, Vulnerabilities and Consequences, Beyza Unal and Patricia Lewis, January 2018, International Security Department, Chatham House, last retrieved on 2 August 2018 at https://www.chathamhouse.org/sites/default/files/publications/research/2018-01-11-cybersecurity-nuclear-weapons-unal-lewis-final.pdf

[75] N. Perlroth, Hackers are targeting nuclear facilities, 6 July 2017, last retrieved on 1 June 2018 at https://www.nytimes.com/2017/07/06/technology/nuclear-plant-hack-report.html?_r=0

sometimes accessible through journalists' investigations, memoirs of former government officials, scientists, security personnel, reports on funding, manufacture, contracts, etc.

The four recent reports, published by Chatham House[76], GAO[77], RAND[78] and NTI[79], review the current status of weapon systems cybersecurity, analyse the threat of cyberattacks against nuclear weapon systems and N3C and testify for the need of urgent actions at all levels of their life cycle. The attacks scenarios analyzed by RAND, and in part by NTI, show the impact of asymmetric cyberwarfare on the current strategic stability and how this affects the so-called "rational deterrence theory". The reports also agree, that a successful cyberattack on nuclear weapons or related systems (e.g., nuclear planning systems, early warning systems, communication systems, delivery systems, etc.) is possible and could lead to catastrophic consequences.

The 2016 NTI Report "Outpacing the Cyber Threat: Priorities for Cybersecurity at Nuclear Facilities" states that "terrorists and other hackers today may not have the cyber skills to facilitate the theft of nuclear bomb-making materials, but they will certainly make every effort. They could use stolen materials to detonate a bomb in any country in the world. They could sabotage systems to cause the release of dangerous levels of radiation that would extend beyond state borders. Or they could hold a facility hostage until their sinister demands were met. Beyond the unthinkable potential human toll, a serious cybersecurity breach would profoundly shake global confidence in civilian nuclear power generation."[80]

The report affirms that "there's no doubt that nuclear facility operators and regulators are aware of the threat. Unfortunately, many of the traditional methods of cyber defence at nuclear facilities—including firewalls, antivirus technology, and air gaps—are no longer enough to match today's dynamic threats."[81]

---

[76] Cybersecurity of Nuclear Weapons Systems Threats, Vulnerabilities and Consequences, Beyza Unal and Patricia Lewis, January 2018, International Security Department, Chatham House, last retrieved on 20 October 2018 at https://www.chathamhouse.org/sites/default/files/publications/research/2018-01-11-cybersecurity-nuclear-weapons-unal-lewis-final.pdf

[77] Weapon Systems Cybersecurity: DOD Just Beginning to Grapple with Scale of Vulnerabilities, 9 October 2018, last retrieved on 20 October 2018 at https://www.gao.gov/mobile/products/GAO-19-128?utm_source=onepager&utm_medium=email&utm_campaign=email_cnsa

[78] How might AI affect the risk of nuclear war, by Edward Geist, Andrew J. Lohn, RAND Report, April 2018, last retrieved on 20 October 2018 at https://www.rand.org/pubs/perspectives/PE296.html,

[79] Nuclear weapons in the new cyber age: Report of the cyber-nuclear weapons study group. September 2018, Page O. Stoutland, PhD and Samantha Pitts-Kiefer, 2016, last retrieved on 22 October 2018 at https://www.nti.org/media/documents/Cyber_report_finalsmall.pdf

[80] A. van Dine and P. Stoutland, Outpacing cyber threats, NTI, last retrieved on 19 October 2018 at https://www.nti.org/media/documents/NTI_CyberThreats__FINAL.pdf

[81] ibid

In 2018 the new NTI report "Nuclear Weapons in the New Cyber Age: Report of the Cyber-Nuclear Weapons Study Group",[82] addresses cyber risks to the nuclear weapons systems and offers recommendations developed by a group of high-level former and retired government officials, military leaders, and experts in nuclear systems, nuclear policy, and cyber threats.

Nuclear Weapons in the New Cyber Age: Report of the Cyber-Nuclear Weapons Study Group

NUCLEAR INDUSTRY

Nuclear facilities and the industry in general, are the first line of defence when addressing cyber threats. In the near future, industry should:

Apply lessons learned from industry experiences with safety and physical security to institutionalize and promote ongoing improvements in cybersecurity at nuclear facilities;

Initiate the development of active defence capabilities at the facility level, including perhaps developing mutual-aid agreements or other cross-industry resources to allow facilities to access needed skills;

Work to reduce system complexity at nuclear facilities by characterizing systems, identifying excess functionalities and removing them where possible, as well as working with vendors to develop non-digital systems and secure-by-design products where appropriate;

Support the cybersecurity efforts of relevant organizations, including the IAEA, the WNA, WANO, and INPO in an effort to continue the international dialogue and contribute to key research and development necessary to improve cybersecurity; and

Provide training opportunities and assistance to boost human capacity across the cyber-nuclear field, especially in countries with new or expanding civilian nuclear energy programs.

INTERNATIONAL ORGANIZATIONS

The magnitude of the threat can overwhelm already overtaxed governments and can strain limited resources. International organizations can help lessen this burden. In the short term, international organizations should:

Support, through international dialogue, provision of guidance and training to governments and facilities, and definition of relevant best practices, international cooperation and an expanded focus on cybersecurity at nuclear facilities;

Facilitate sharing of threat information where possible and appropriate;

Provide platforms for discussing and developing solutions for reducing complexity; and **Foster innovation** and continue to think creatively about how to mitigate the threat and recruit a variety of voices and perspectives to join the conversation.

The 2018 NTI Report[83]

---

[82] Nuclear Weapons in the New Cyber Age: Report of the Cyber-Nuclear Weapons Study Group, NTI Report, 26 September 2018, last retrieved on 30 September 2018 at https://www.nti.org/analysis/reports/nuclear-weapons-cyber-age/
[83] Ibid

Futter states that "Nuclear command and control systems have always been vulnerable to outside interference, attack and possible sabotage. The declassified background confirms miscalculations, accidents and near misses (many of which were caused by computers and electronic systems) …. primarily due to the central challenge of balancing two separate but co-constitutive nuclear requirements: the need for positive control (ensuring that weapons will work and can be used under all circumstances) and the need for negative control (ensuring that weapons are never used by accident or by unauthorized actors)".[84]

Nuclear weapons have been developed, produced and stockpiled, but not used for conflict resolution since 1945. However, even their storage and maintenance have been accompanied by the constant danger of catastrophic consequences. The 2014 Chatham House Report "Too Close for Comfort Cases of Near Nuclear Use and Options for Policy" reveals that "Evidence from many declassified documents, testimonies and interviews (civil society and advocacy organizations and journalists, coupled with a move towards declassification and freedom of information in some countries, a number of cases of near-nuclear weapons use have come to light) suggests that the world has, indeed, been lucky, given the number of instances in which nuclear weapons were nearly used inadvertently as a result of miscalculation or error".[85] It provides a list of the known near-nuclear events (see Figure 2.3). This list can be added to with near nuclear incidents from the NTI web-site.[86]

Near-Nuclear Events Highlights

In 1980, the warning systems showed missiles headed for the United States. In the minutes remaining before the President would have had to order a retaliatory strike, the warning was determined to be a false alarm caused by a faulty computer chip.[87]

In **1991,** it was feared that a group of Dutch hackers who broke into the US military networks were searching for nuclear secrets and missile data to sell to Iraqi leader Saddam Hussein prior to Operation Desert Storm.[88]

---

[84] A. Futter, Cyber Threats and Nuclear Weapons, Royal United Services Institute for Defence and Security Studies, July 2016, last retrieved on 20 October 2018 at https://rusi.org/sites/default/files/cyber_threats_and_nuclear_combined.1.pdf

[85] Too Close for Comfort Cases of Near Nuclear Use and Options for Policy, Chatham House Report, Patricia Lewis, Heather Williams, Benoît Pelopidas and Sasan Aghlani April 2014, last retrieved on 2 September 2018 at https://www.chathamhouse.org/sites/default/files/field/field_document/20140428TooCloseforComfortNuclearUseLewisWilliamsPelopidasAghlani.pdf

[86] References for Cyber Incidents at Nuclear Facilities, last retrieved on 22 October 2018 at https://www.nti.org/analysis/tools/table/133/

[87] The 3 A.M. Phone Call, False Warnings of Soviet Missile Attacks during 1979-80 Led to Alert Actions for U.S. Strategic Forces, 1 March 2012, last retrieved on 22 October 2018 at https://nsarchive2.gwu.edu/nukevault/ebb371/

[88] Dorothy Denning, 1999, Information Warfare and Security, Addison-Wesley, Reading, MA.

In **1998,** the Cox Report revealed that China had stolen a considerable cache of highly sensitive secrets over a number of years from the US, particularly those relating to the W88 thermonuclear warhead design.[89] Matthew McKinzie later remarked that it was an 'unprecedented act of espionage … The espionage in the Manhattan Project [would] pale in comparison.'[90] This became known as Kindred Spirit.[91]

Later that year, an American teenage hacker broke into the India's Bhabha Atomic Research Centre (BARC) and downloaded passwords and emails.[92]

In **1999**, the Moonlight Maze attack, believed to have emanated from Russia, was revealed to have stolen thousands of files and other pieces of sensitive information, and to have infiltrated deep into the Pentagon and other US government departments.[93]

In **August 2007**, six US nuclear-armed cruise missiles were missing for 36 hours. They were mistakenly placed under the wings of a B-52 and were not guarded according to protocol during a subsequent flight from Minot Air Force Base in North Dakota to Barksdale, Louisiana. Had the plane experienced any problem in flight, the crew would not have known to follow the proper emergency procedures with nuclear weapons on board.[94]

> Today's strategic nuclear triad of the US "consists of: submarines (SSBNs) armed with submarine-launched ballistic missiles (SLBM); land-based intercontinental ballistic missiles (ICBM); and strategic bombers carrying gravity bombs and air-launched cruise missiles (ALCMs)".[95]

In **February 2009**, a collision between the nuclear armed UK HMS Vanguard and French Le Triomphant demonstrates the risk posed by

---

[89] Select Committee US House of Representatives, 'Report of the Select Committee on US National Security and Military/Commercial Concerns with the Republic of China', 25 May 1999, Chapter 2, last retrieved on 6 June 2018.

[90] Vernon Loeb and Walter Pincus, 'Los Alamos Security Breach Confirmed', Washington Post, 29 April 1999.

[91] Dan Stober and Ian Hoffman, 1989, A Convenient Spy: Wen Ho Lee and the Politics of Nuclear Espionage Doubleday, New York NY, Notra Trulock, 2002 Code Name Kindred Spirit: Inside The Chinese Nuclear Espionage Scandal, Encounter Books, San Francisco; Shirley Kan, 'China: Suspected Acquisition of U.S. Nuclear Weapon Secrets', US Congressional Research Service, RL30143, 1 February 2006.

[92] Adam Penenberg, 'Hacking Bhabha', Forbes, 16 November 1998.

[93] Adam Elkus, 'Moonlight Maze' in Jason Healey (ed.), 2013. 'A Fierce Domain: Conflict in Cyberspace, 1986 to 2012', Cyber Conflict Studies Association, p. 155.

[94] The quotes for the account of this particular accident are taken from the unclassified account available in the February 2008 report from the Defense Science Board Permanent Task Force on Nuclear Weapons Surety, Report on the Unauthorized Movement of Nuclear Weapons.

[95] Nuclear posture review, February 2018, Office of the Secretary of Defence, last retrieved on 2 September 2018 at https://media.defense.gov/2018/Feb/02/2001872886/-1/-1/1/2018-NUCLEAR-POSTURE-REVIEW-FINAL-REPORT.PDF

Incidents of near nuclear use

Date	Incident	States involved	Cause
October 1962	Operation Anadyr	Soviet Union	Miscommunication
27 October 1962	British nuclear forces during the Cuban missile crisis	United Kingdom	Conflict escalation
27 October 1962	Black Saturday	United States	Conflict escalation and miscommunication
22 November 1962	Penkovsky false warning	Soviet Union	Espionage
October 1973	1973 Arab–Israeli war	Israel	Conflict escalation
9 November 1979	NORAD: Exercise tape mistaken for reality	United States	Exercise scenario tape causes nuclear alert
3 June 1980	NORAD: Faulty computer chip	United States	Faulty computer chip
25 September 1983	Serpukhov-15	Soviet Union	Technical error
7–11 November 1983	Able Archer-83	Soviet Union, United States	Misperception of military training exercise
18–21 August 1991	Failed coup	Soviet Union	Loss of command and control structure
25 January 1995	Black Brant scare	Russia	Mistaken identity of research rocket launch
May–June 1999	Kargil crisis	India, Pakistan	Conflict escalation
December 2001–October 2002	Kashmir standoff	India, Pakistan	Conflict escalation

vi | Chatham House

Figure 2.3: Incidents of near nuclear use

Source: Too Close for Comfort Cases of Near Nuclear Use and Options for Policy, Chatham House Report, 2014

current attitudes towards intelligence and transparency on nuclear weapons issues. The two nuclear powered, ballistic missile-carrying submarines (SSBNs), collided in the Atlantic Ocean.[96]

In April 2009, the Secretary of State for Defence Bob Ainsworth was asked to list all collisions that involved UK nuclear submarines with other vessels, as well as the grounding of UK nuclear submarines since 1979. Though stating that the data on the incidents that might have occurred between 1979 and 1988 were not held centrally, Ainsworth cited 14 other incidents that occurred between 1988 and 2008.[97]

In 2010, 50 nuclear-armed Minuteman missiles from the underground silos in Wyoming mysteriously disappeared from their launching crews' monitors for nearly an hour. The crews could not have fired the missiles on presidential orders or discerned whether an enemy was trying to launch them. It was unclear, whether it was a technical malfunction or a security breach. As per the initial readings, someone had put all 50 missiles into countdown to launch. The missiles were designed to fire instantly as soon as they received a short stream of computer code, and they are indifferent about the code's source.[98]

In July 2012, the Y-12 facility for handling, processing and storing weapons-grade uranium has been temporarily shut after anti-nuclear activists breached security fences.[99]

In 2013, the security vulnerabilities were revealed on L'Ile Longue, the heart of nuclear defence, which hosts France's four ballistic missile submarines.[100]

Long-term concerns regarding competent personnel and strategic organizational change were highlighted by the 2012–13 report by the Defence Nuclear Safety Regulator to the UK Ministry of Defence.[101]

In April 2016, reports surfaced that the Gundremmingen Nuclear Power Plant in Bavaria was infected with malware. The discovery was

---

[96] 'Nuclear subs collide in Atlantic', BBC News, 16 February 2009, http://news.bbc.co.uk/1/hi/7892294.stm; UK Ministry of Defence, last retrieved on 2 September 2018 at http://nuclearinfo.org/sites/default/files/Submarine%20collision%20FOI%20release%20270213.pdf

[97] 'Commons Hansard Written Answers Text for 2 April 2009', Column 1396W, last retrieved on 3 September 2018 at http://www.publications.parliament.uk/pa/cm200809/cmhansrd/cm090402/text/90402w0024.htm.

[98] https://www.nytimes.com/2017/03/14/opinion/why-our-nuclear-weapons-can-be-hacked.html

[99] M. Hosenball, 'Oak Ridge uranium plant shut after protesters breach 4 fences, reach building', Reuters, 2 August 2012, last retrieved on 2 October 2018 at http://uk.reuters.com/article/2012/08/02/us-usa-securtity-nuclear-idUSBRE8711LG20120802

[100] H. Chambonnière, Hervé, 'Ile Longue. Les incroyables failles dans la sécurité' [Ile Longue. Incredible security breaches], Le Telegramme de Brest, 11 June 2013, last retrieved on 13 November 2018 at http://www.letelegramme.fr/ig/generales/fait-du-jour/ile-longue-des-failles-dans-la-securite-11-06-2013-2132250.php

[101] Defence Nuclear Safety Regulator Annual Report 2012/2013, last retrieved on 2 September 2018 at https://www.gov.uk/government/uploads/system/uploads/attachment_data/file/212708/dnsr annual_report_2012_2013.pdf

made in the plant's B unit, in a computer system that had been retrofitted in 2008 with data-visualization software accompanying equipment for moving nuclear fuel rods. Viruses had also infected 18 removable data drives associated with computers not connected to the plant's operating systems. There was no apparent damage. This statement raises questions about how the "isolated" plant became infected and why the malware went undetected for so long.[102]

In June 2016, hackers used a spear-phishing attack to steal research and personal data from the University of Toyama Hydrogen Isotope Research Center. This research centre is a world leader in research into tritium, a radioactive isotope of hydrogen that serves as fuel for controlled nuclear fusion and is an integral part of hydrogen bombs.[103]

Among the cases of cyber espionage related to nuclear weapons the following might be named:

In February 2011, the Zeus was discovered, an information-stealing Trojan aimed at contractors involved in building the UK Trident nuclear-armed submarine force.[104]

In May 2011, Iran was accused of hacking the International Atomic Energy Agency (IAEA), looking for secrets regarding the monitoring of its nuclear programme.[105]

In August 2011, the Shady RAT malware targeted US government agencies, defence contractors and numerous high-technology companies.[106]

In November 2012, the group Anonymous claimed to have hacked IAEA and threatened to release 'highly sensitive data' on the Israeli nuclear programme that they had allegedly seized.[107]

In 2013, hackers believed to be from the group Deep Panda (linked with the Chinese PLA) targeted the computers of the US nuclear researchers directly.[108]

Hackers have also sought to attack nuclear-related systems, e.g. the US and Israeli ballistic missile defence (BMD) programmes, and are suspected of stealing classified data from those systems, too.[109]

---

[102] The 2018 NTI Report Outpacing Cyber Threats: Priorities for Cybersecurity at Nuclear Facilities, last retrieved on 21 October 2018 at https://www.nti.org/media/documents/NTI_CyberThreats__FINAL.pdf

[103] Ibid

[104] Richard Norton-Taylor, 'Chinese Cyber-Spies Penetrate Foreign Office Computers', The Guardian, 4 February 2011

[105] David Crawford, 'UN Probes Iran Hacking of Inspectors', Wall Street Journal, 19 May 2011

[106] Hagestad, 21st Century Chinese Cyberwarfare, p. 12.

[107] Michael Kelley, 'Anonymous Hacks Top Nuclear Watchdog Again to Force Investigation of Israel', Business Insider, 3 December 2012.

[108] Russia Today, 'US Nuclear Weapons Researchers Targeted with Internet Explorer Virus', 7 May 2013

[109] Global Security Newswire, 'Chinese Hacking Targets US Missile Defense Designs', 28 May 2013, last retrieved on 21 October 2018 at https://www.nti.org/gsn/article/chinese-hacking-targets-us-missile-defense-designs/

Risk Management

The risks in nuclear weapons development, production, storage, transfer and disposal are conditioned by the increasing complexity of the weapons themselves, including through computerization and digitization, and consequent increase in vulnerabilities. In addition, any nuclear facility is exposed to the risk of routine incidents and even accidents, including through environmental factors.

These high-level risks raise significant doubts as to the reliability and integrity of nuclear weapons systems, regarding their ability to: a) launch a weapon; b) prevent an inadvertent launch; c) maintain command and control of all military systems; d) transmit information and other communications; and e) the maintenance and reliability of such systems.[110]

The 2018 GAO Report to the Committee on Armed Services warns that "the weapon systems are increasingly dependent on software and IT to achieve their intended performance. The amount of software in today's weapon systems is growing exponentially and is embedded in numerous technologically complex subsystems, which includes hardware and a variety of IT components" [111] (See Figure 2.4). Communications as well as the transfer and storage of data are key targets for cyber attackers.

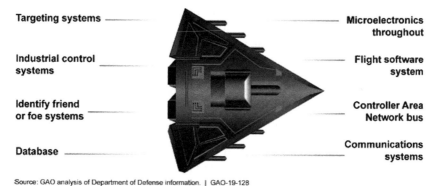

Targeting systems — Microelectronics throughout

Industrial control systems — Flight software system

Identify friend or foe systems — Controller Area Network bus

Database — Communications systems

Source: GAO analysis of Department of Defense information. | GAO-19-128

Figure 2.4: Software in weapon systems

The International Security Department at Chatham House identified several areas within the nuclear weapons systems that could be potentially vulnerable to cyberattacks:[112]

---

[110] Ibid

[111] GAO Report to the Committee on Armed services, U.S. Senate, Weapon Systems Cybersecurity, October 2018, last retrieved on 26 November 2018 at https://www.gao.gov/assets/700/694913.pdf

[112] Cybersecurity of Nuclear Weapons Systems Threats, Vulnerabilities and Consequences, Beyza Unal and Patricia Lewis, January 2018, International Security Department, Chatham House, last retrieved on 1 September 2018 at https://www.chathamhouse.org/sites/default/files/publications/research/2018-01-11-cybersecurity-nuclear-weapons-unal-lewis-final.pdf

- Communications between command and control centres;
- Communications from command stations to missile platforms and missiles;
- Telemetry data from missiles to ground- and space-based command and control assets;
- Analytical centres for gathering and interpreting long-term and real-time intelligence;
- Cyber technologies in transport;
- Cyber technologies in laboratories and assembly facilities;
- Pre-launch targeting information for upload;
- Real-time targeting information from space-based systems including positional, navigational and timing data from global navigational systems;
- Real-time weather information from space-, air-, and ground-based sensors;
- Positioning data for launch platforms (e.g. submarines);
- Real-time targeting information from ground stations;
- Communications between allied command centres.

Compared to all the above technological threats, the human factor remains the leading vulnerability. Human reactions to the signals of an upcoming nuclear attack, even though falsely alerted by a cyberattack, can be to launch an immediate response through adequate nuclear means, such as ballistic nuclear missiles.

With the growing number of deception techniques, the issue of human decision making and time to analyse the reliability of the threat signal is one of major challenges faced. For instance, "Russia works on a new spoofing device that can imitate jets, rockets or a missile attack and thus fool defence systems".[113]

Increased engagement of the private sector poses additional risks. Unal and Lewis[114] state that "Many aspects of nuclear weapons development and systems management are privatized in the US and in the UK, potentially introducing a number of private-sector supply chain vulnerabilities."

To summarize the above, it is worth concluding that vulnerabilities, and the possibility to embed malicious cyber arms, exist through the whole

---

[113] S. Sukhankin, 2017. 'Russian Capabilities in Electronic Warfare: Plans, Achievements and Expectations', Real Clear Defence, 20 July 2017, last retrieved on 6 December 2018 at https://www.realcleardefense.com/articles/2017/07/20/russian_capabilities_in_electronic_warfare_111852.html

[114] Cybersecurity of Nuclear Weapons Systems Threats, Vulnerabilities and Consequences, Beyza Unal and Patricia Lewis, January 2018, International Security Department, Chatham House, last retrieved on 1 September 2018 at https://www.chathamhouse.org/sites/default/files/publications/research/2018-01-11-cybersecurity-nuclear-weapons-unal-lewis-final.pdf

life cycle of nuclear weapons—from design and production to supply chain and maintenance. Vulnerabilities exist in warning, communication, security systems, radars, satellites, remote control systems, etc. The overall sophistication of the cyberspace environment—computers, networks, AI-enhanced technologies, along with the consequent sophistication of attacks—should ensure adequate regulation of the still present and essentially needed human control for authorization to use weapons.

> "We found that from 2012 to 2017, DOD testers routinely found mission-critical cyber vulnerabilities in nearly all weapon systems that were under development. Using relatively simple tools and techniques, testers were able to take control of these systems and largely operate undetected. In some cases, system operators were unable to effectively respond to the hacks. Furthermore, DOD does not know the full scale of its weapon system vulnerabilities because, for a number of reasons, tests were limited in scope and sophistication".[115]
>
> *GAO Report to the Committee on Armed Services, 2018*

Weapon systems have a wide variety of interfaces, some of which are not obvious, that could be used as pathways for adversaries to access systems, as is shown in Figure 2.5.[116]

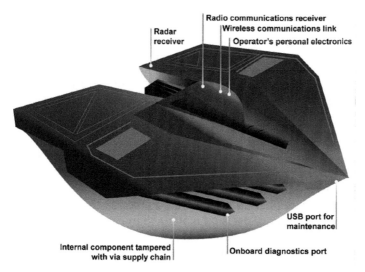

Figure 2.5: Weapons interfaces that can be used as pathways to access the system

Source: GAO Report, GAO-19-128

---

[115] GAO Report to the Committee on Armed Services, U.S. Senate, Weapon Systems Cybersecurity, October 2018, last retrieved on 26 November 2018 at https://www.gao.gov/assets/700/694913.pdf

[116] Ibid

To build a resilient weapon system, it is essential, that the following measures are introduced, among many others:

- Regulation of acquisition and certified producers
- Development of system-specific cyber requirements
- Routine verification and cybersecurity assessments
- Testing with incorporated cybersecurity process for malfunctions and vulnerabilities, updates and security patches
- Improved verification of threats and adjusted response time
- Deployed redundancy measures
- Creation of cyber incident hotline

Implementing cyber security measures in weapon systems is a challenging task due to their complexity, and interconnectedness with other controlling and monitoring systems, high security levels, multiple approval steps, etc. For example, a security update can be applied within a short timeframe after its development, but it might take weeks or even months to verify the compatibility and stability of the system with the updated software. Some systems can only be updated or upgraded using specific equipment from the original manufacturer or at the maintenance facility or location.

> The interaction between cyber operations and nuclear weapons is a complex problem. The complexity and danger of interaction is only partly a function of technology but also a function of significant political, economic, and strategic differences associated with the use of cyber and nuclear capabilities. There are many potential vulnerabilities in the expanding attack surface of modern N3C, but the challenges associated with exploiting those vulnerabilities under actual operational and political conditions are nontrivial, to put it mildly. As a result, not every state will have the same capability to conduct cyber operations against enemy N3C, and non-state actors are unlikely to have much capability at all.
>
> *Jon Lindsay, 20 June 2019* [117]

For as long as nuclear weapons exist, the risk of an inadvertent, accidental or deliberate detonation remains. Until their elimination, vigilance and prudent decision making in nuclear policies are therefore of the utmost priority. Responses that policy-makers and the military should consider include buying time for decision-making, particularly in crises; developing trust and confidence building measures; refraining from large-scale military exercises during times of heightened tension; involving a

---

[117] Jon Lindsay, "Cyber Operations and Nuclear Weapons", NAPSNet Special Reports, June 20, 2019, https://nautilus.org/napsnet/napsnet-special-reports/cyber-operations-and-nuclear-weapons/

wider set of decision-makers in times of crisis; and improving awareness and training on the effects of nuclear weapons.[118]

The 2014 Chatham House Report "Too Close for Comfort Cases of Near Nuclear Use and Options for Policy" concludes: "Institutionalizing cybersecurity at nuclear facilities, implementing active defence strategies, and minimizing complexity would address many of the serious vulnerabilities the world faces today, and investing in transformative research and development will lay the groundwork for an even more secure future. Governments, industry, and international organizations all have a role to play in addressing and outpacing this threat. The risk is too great to accept the status quo".[119]

2.6 Cyber Research and Defence Modelling

From the "scientific" creation and analysis of the first virus, cyber security researchers have been providing a valuable input into developing countermeasures against cyber threats. They not only develop protection mechanisms against cyber arms, but also model future challenges and defence models, including for critical infrastructure and CBRNe.

2.6.1 Research Highlights and Approaches

Cyber research is conducted in government research centres, universities, private corporations and start-ups. They set proactive research goals, simulate attack scenarios, develop defence options and inform policy makers on the upcoming threats and challenges.

Highlights

In 2013, after reports on the growing number of industrial cyber security attacks, and along with government initiatives such as the formation of the Cyber Security Information Sharing Partnership (CISP), Siemens put the protection of industrial control systems firmly in the spotlight.[120]

---

[118] Too Close for Comfort Cases of Near Nuclear Use and Options for Policy, Chatham House Report, Patricia Lewis, Heather Williams, Benoît Pelopidas and Sasan Aghlani, April 2014, last retrieved on 12 November 2018 at https://www.chathamhouse.org/sites/default/files/field/field_document/20140428TooCloseforComfortNuclearUseLewisWilliamsPelopidasAghlani.pdf

[119] The 2018 NTI Report Outpacing Cyber Threats: Priorities for Cybersecurity at Nuclear Facilities, last retrieved on 19 October 2018 at https://www.nti.org/media/documents/NTI_CyberThreats__FINAL.pdf

[120] Minimizing cyber security risk for the Chemical sector, Siemens, 29 October 2014, last retrieved on 16 November 2018 at http://w3.siemens.co.uk/home/uk/en/iadt/news/pages/minimising-cyber-security-risk-for-the-chemical-sector.aspx

In 2015, security researchers at Black Hat[121] USA described a proof-of-concept worm that targets weaknesses within automated industrial control systems used to manage critical infrastructure and manufacturing. The worm, according to OpenSource Security, has the capability to autonomously search for and spread between networked PLCs. They also developed mitigation techniques and protection features.[122]

In 2016, the Pittsburgh-based team ForAllSecure's Mayhem Cyber Reasoning System (CRS) took first place at the Cyber Grand Challenge finals, beating out six other computers. The Mayhem CRS is now on display at the Smithsonian's National Museum of American History.

In 2017, cybersecurity researchers at the Georgia Institute of Technology[123] developed a new form of ransomware that was able to take over control of a simulated water treatment plant. After gaining access, the researchers were able to command PLCs to shut the valves, increase the amount of chlorine added to water, and display false readings. This event proved once again (after the Stuxnet event) that critical infrastructure are vulnerable to cyberattacks, and there is no ultimate defence against the false sense of security induced by false values and indicators on the screens and displays.

In 2018, the Imperial College London described a deep learning-powered cyberattack framework for Industrial Networks. The framework reportedly allows an attacker to conduct covert cyberattacks with minimal prior knowledge of the target ICS. As per the report, the findings motivate greater attention to this area by the security community as they demonstrate that the currently assumed barriers for the successful execution of such attacks are relaxed. Much like with early malware, such attention is likely to inspire the development of innovative security measures by providing an understanding of the limitations of the current detection implementations and will inform security by design considerations for those planning future ICS.[124]

Cyber Security Centres of Excellence certified and funded by governments appeared globally, serving as a connection point between government defence agencies and academia. Their joint projects facilitate research in the most critical areas of cyber defence.

---

[121] T. Spring, PLC-Blaster Worm Targets Industrial Control Systems, 5 August 2016, last retrieved on 15 August 2018 at https://www.blackhat.com/docs/asia-16/materials/asia-16-Spenneberg-PLC-Blaster-A-Worm-Living-Solely-In-The-PLC-wp.pdf

[122] T. Spring, PLC-Blaster Worm Targets Industrial Control Systems, 5 August 2016, last retrieved on 15 August 2018 at https://threatpost.com/plc-blaster-worm-targets-industrial-control-systems/119696/

[123] Simulated Ransomware Attack Shows Vulnerability of Industrial Controls, 13 February 2017, last retrieved on 12 July 2018 at http://www.rh.gatech.edu/news/587359/simulated-ransomware-attack-shows-vulnerability-industrial-controls

[124] C. Feng, T. Li, Z. Zhu, D. Chana, A Deep Learning-based Framework for Conduction Stealthy Attacks in Industrial Control Systems, Imperial College London, 2018, last retrieved on 18 October 2018 at https://arxiv.org/pdf/1709.06397.pdf

> **In 2012**, Imperial College London, Royal Holloway, and Oxford University, were among the first recognized in the UK as academic centres of excellence in cybersecurity research. This recognition is given by the National Cyber Security Centre, the UK's authority on cybersecurity.
>
> "We are now living in a world that is increasingly fragmented and more hostile. It has never been more important for the UK to be prepared to defend itself against a range of cyber threats. Academic Centres of Excellence, like Imperial's, play a really important role in this process and it means that we can work with the UK government and industry to help shape the research landscape in cyber security."[125]
>
> *Dr Lupu, Associate Director at the Institute for Security Science and Technology*

Countries joined their efforts under bilateral and global treaties, contributing to the development of the best cybersecurity solutions and legal frameworks for cyber space.

> **NATO Cooperative Cyber Defence Centre of Excellence**
>
> The NATO Cooperative Cyber Defence Centre of Excellence is a NATO-accredited knowledge hub, research institution, and training and exercise facility. The Tallinn-based international military organization focuses on interdisciplinary applied research, consultations, trainings and exercises in the field of cybersecurity.
>
> NATO CCD COE is the home of the Tallinn Manual 2.0 on the International Law Applicable to Cyber Operations.
>
> The Centre organises the world's largest and most complex international technical cyber defence exercise Locked Shields and the annual conference on cyber conflict, CyCon.
>
> The Centre is a multinational and interdisciplinary hub of cyber defence expertise, uniting practitioners from 20 nations. The heart of the Centre is a diverse group of experts: researchers, analysts, trainers and educators. The mix of military, government and industry backgrounds means the NATO CCD COE provides a unique 360-degree approach to cyber defence. The organization supports its member nations and NATO with cyber defence expertise in the fields of technology, strategy, operations, and law.
>
> As of 2017, Belgium, the Czech Republic, Estonia, France, Germany, Greece, Hungary, Italy, Latvia, Lithuania, the Netherlands, Poland, Slovakia, Spain, Turkey, the United Kingdom and the United States are Sponsoring Nations of the NATO Cooperative Cyber Defence Centre of Excellence. Austria and Finland have become Contributing Participants, and Sweden is well on its way to following suit. The Centre is staffed and financed by member nations and, as such, it is not part of NATO's military command or force structure.[126]

Academia research capacities join with businesses providing innovative solutions, developing cyber science and showing best practices in the management of high technology companies. One of the

---

[125] Imperial recognized as a centre of excellence in cyber security research, 3 April 2017, last retrieved on 19 October 2018 at https://www.imperial.ac.uk/news/178042/imperial-recognised-centre-excellence-cyber-security/

[126] Tallinn Manual 2.0 on the International Law Applicable to Cyber Operations, NATO Cooperative Cyber Defence Centre of Excellence, last retrieved on 19 August 2018 at https://ccdcoe.org/sites/default/files/documents/CCDCOE_Tallinn_Manual_Onepager_web.pdf

best examples of this efficiency is Akamai Technologies, Inc., the world's largest cloud delivery platform. Its history dates back to 1998, when the founders, MIT professors and students, obtained an exclusive license to certain intellectual property from MIT. Soon after that, experienced Internet business professionals joined the team and now amount to more than 1,600 experts.

"Currently, Akamai provides a defensive shield built to protect websites, mobile infrastructure, and API-driven requests. Via 24/7 monitoring, it collects and analyses terabytes of attack data, billions of bot requests, and hundreds of millions of IP addresses to solidify defences. Akamai's platform is unparalleled in scale with over 200,000 servers across 130 countries, giving customers ... threat protection."[127]

The Akamai *Summer 2018 State of the Internet / Security: Web Attack* report highlights:
- A 16 percent increase in the number of DDoS attacks recorded since last year.
- The largest DDoS attack of the year set a new record at 1.35 Tbps by using the *memcached* reflector attack.
- Researchers identified a 4 percent increase in reflection-based DDoS attacks since last year.
- A 38 percent increase in application-layer attacks such as SQL injection or cross-site scripting.
- In April, the Dutch National High Tech Crime Unit took down a malicious DDoS-for-hire website with 136,000 users.[128]

Project Maven of the US Department of Defense is an example of a multi-million government and business collaboration. Using TensorFlow-based (Google) Artificial Intelligence systems, it develops algorithms to interpret imagery from drone surveillance collection. The project testifies that AI is actively employed by militaries to interpret data more accurately, and to quickly analyse publicly available information, etc., which in turn contributes to better decision making.[129] It is worth noting that currently business is cautious and sometimes reluctant to implement military projects for military applications of high technologies.

2.6.2 Artificial Intelligence: Role and Influence

One day the AIs are going to look back on us
the same way we look at fossil skeletons on the plains of Africa.
An upright ape living in dust with crude language
and tools, all set for extinction.[130]

---

[127] Akamai, last retrieved on 14 October 2018 at https://www.akamai.com

[128] Akamai State of the Internet / Security Summer 2018: Web Attack Report Shows Hospitality Industry Under Siege From Botnets, 26 June 2018, last retrieved on 14 October 2018 at https://www.akamai.com/us/en/about/news/press/2018-press/akamai-releases-summer-2018-state-of-the-internet-security-report.jsp

[129] S. Gibbs, Google's AI is being used by US military drone programme, The Guardian, 7 March 2018, last retrieved on 20 May 2018 at https://www.theguardian.com/technology/2018/mar/07/google-ai-us-department-of-defense-military-drone-project-maven-tensorflow

[130] Alex Garland, Ex Machina, a sci-fi thriller, 2015

After a series of AI research winter periods, modern technological development allowed further advancements in AI and machine learning initiated half a century ago. Supported by government and private investments, research in AI development is skyrocketing. Among the leading AI companies there are AIBrain, Amazon, Anki, Apple, Banjo, CloudMinds, Deepmind, Facebook, Google, H20, IBM, iCarbonX, Iris AI, Microsoft, Next IT, Nvidia, OpenAI[131], which create platforms for the worldwide AI applications.

The high volume of investments is justified by the forecast that AI could contribute up to 15.7 trillion USD to the global economy in 2030, more than the current output of China and India combined. Of this, 6.6 trillion USD is likely to come from increased productivity and 9.1 trillion USD is likely to come from consumption side effects.[132]

There is still no commonly agreed wording, even among computer scientists and engineers, but a general definition of AI is the capability of a computer system to perform tasks that normally require human intelligence. In computer science AI research is defined as the study of intelligent agents: "any device that perceives its environment and takes actions that maximize its chance of successfully achieving its goals."[133]

> "An intelligent agent (IA) is an autonomous entity which observes through sensors and acts upon an environment using actuators (i.e., it is an agent) and directs its activity towards achieving goals. Intelligent agents may also learn or use knowledge to achieve their goals".[134]
>
> **Automated intelligence:** Automation of manual/cognitive and routine/nonroutine tasks.
> **Assisted intelligence:** Helping people to perform tasks faster and better.
> **Augmented intelligence:** Helping people to make better decisions.
> **Autonomous intelligence:** Automating decision-making processes without human intervention.[135]

AI capabilities[136] are becoming more powerful and now include reasoning, machine learning, language and data set processing, object

---

[131] Andy Patrizio, Top 25 Artificial Intelligence Companies, 10 April 2018, last retrieved on 15 August 2018 at https://www.datamation.com/applications/top-25-artificial-intelligence-companies.html

[132] Sizing the prize: What's the real value of AI for your business and how can you capitalise?, PWC, 2017, last retrieved on 20 May 2018 at https://www.pwc.com/gx/en/issues/analytics/assets/pwc-ai-analysis-sizing-the-prize-report.pdf

[133] D. Poole, A. Mackworth and R. Goebel, 1998. Computational Intelligence: A Logical Approach. Oxford University Press. New York.

[134] S. Russell and P. Norvig, 2003. Artificial Intelligence: A Modern Approach (2nd ed.), Prentice Hall, Upper Saddle River, New Jersey, Ch. 2.

[135] Sizing the prize: What's the real value of AI for your business and how can you capitalise?, PWC, 2017, last retrieved on 20 May 2018 at https://www.pwc.com/gx/en/issues/analytics/assets/pwc-ai-analysis-sizing-the-prize-report.pdf

[136] S. Russell and P. Norvig, 2003. Artificial Intelligence: A Modern Approach (2nd ed.), Upper Saddle River, New Jersey: Prentice Hall, Ch. 2.

perception, information storage and retrieval, and speech and handwriting recognition. They are used in self-driving cars, automated translation, search engines, game-playing robots; they produce voice commands and provide AI-powered server configurations. All this enhances and at the same time endangers cybersecurity.

The AI field of study draws upon computer science, mathematics, psychology, linguistics, philosophy and other fields. Following the development of artificial general intelligence, some analysts[137] [138] forecast the emergence of artificial super-intelligence, a type of AI that far surpasses human intellect and abilities in nearly all areas. Figure 2.6 shows predicted AI capabilities development in the far future.

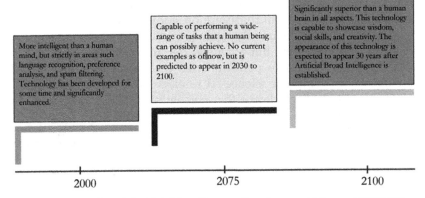

Figure 2.6: Predicted Artificial Intelligence Development (*Source*: EIU 2016)

In connection with the above, three AI generations are identified as Artificial Narrow Intelligence—machine intelligence that equals or exceeds human intelligence for specific tasks—currently used in search engines or the assistants on mobile phones; Artificial General Intelligence—machine intelligence meeting the full range of human performance across any task—controversial, but most experts expect it by the middle of this century; and Artificial Superintelligence: machine intelligence that exceeds human intelligence across any task—predicted to emerge relatively quickly thereafter, although few anticipate this to synthesise real artificial consciousness.[139]

---

[137] J.P. Holdren and M. Smith, Preparing for the Future of Artificial Intelligence White House Executive Office of the President, 2016, last retrieved 29 June 2018 at https://obamawhitehouse.archives.gov/sites/default/files/whitehouse_files/microsites/ostp/NS TC/preparing_for_the_future_of_ai.pdf

[138] T. Urban, The AI Revolution: The Road to Superintelligence Wait But Why, January 2015, last retrieved 29 June 2018 at http://waitbutwhy.com/2015/01/artificial-intelligence-revolution-1.html

[139] Artificial Intelligence and the Future of Defence, Strategic implications for small- and medium-sized force providers, 2017, The Hague Centre for Strategic Studies, last retrieved on 15 August 2018 at https://hcss.nl/sites/default/files/files/reports/Artificial%20Intelligence%20and%20the%20Future%20of%20Defense.pdf

Research for AI real-life applications are done in the area of the narrow AI, which is based on machine learning with the use of large amounts of data. The data used for machine learning can be either supervised—data with associated facts, such as patterns of values or labels—or unsupervised—using a raw data flow that requires the identification of patterns without prior prompting. This includes reinforcement learning, where machine-learning algorithms actively choose and even generate their own training data.

Machine learning tasks are implemented through various learning algorithms, such as artificial neural networks, which consist of the layers of nonlinear transformation node functions, where the output of each layer becomes an input to the next layer in the network. Each layer is highly modular, making it possible to take a layer optimized for one type of data (say, images) and to combine it with other layers for other types of data (e.g., text).

Machine learning tools could be applied in military operations, both in strategic decision making and in tactical applications. AI enhanced computerized systems can implement a vast variety of external and internal operational tasks, from intelligence gathering to managing logistics and implementing administrative duties. In addition, AI can test new equipment.

Models are being developed to forecast AI application and defence. The Artificial Intelligence and the Future of Defence study implemented in 2017 by the Hague Centre for Strategic Studies examined the implications of AI for defence and security organizations (DSOs)[140] and proposed a four defence layers structure of the Armed Forces against the three AI generations (See Table 2.2).

Table 2.2: A four defence layers structure of the armed force against the three AI generations

Levels	Defence layers
1	Armed forces: humans, hardware and software
2	Supporting defence organizations to enable Layer 1 proper functioning—military support entities and the Ministry of Defence
3	All entities across governments responsible for defence and security (a whole-of-government security-oriented approach)
4	Defence and security ecosystem (can be actuated in order to achieve defence and security goals)

Source: The Hague Centre for Strategic Studies

---

140 Ibid

Further on, the study mapped these four layers of 'defence' against the three AI layers. Figure 2.7 shows the options available for small- and medium-sized force providers.

The study concludes that most of the current research on AI and defence focuses on the top-left cell of this table, Artificial Narrow Intelligence. Significant value-for-money opportunities were identified downwards towards the supporting organization, ministries of defence, defence and security ecosystem. Some of the most promising defence and security AI applications are likely to emerge in and be focused on the fourth layer of defence: the defence and security ecosystem layer.

	Artificial Narrow Intelligence	Artificial General Intelligence	Artificial Super Intelligence
'Armed Force'	Monitor frontrunners and acquire opportunistically	Review robustness current force structure	
Ministry of Defense	Identify short-term challenges and opportunities	Identify long-term challenges and opportunities	
Comprehensive	Identify short-term challenges and opportunities	Identify long-term challenges and opportunities	Catalyze dialogue with all stakeholders
Defense and Security Ecosystem	Explore new niches	Explore new niches	Existential challenges and opportunities: fundamental rethink of defense

Figure 2.7: Artificial Intelligence and four layers of defence
Source: The Hague Centre for Strategic Studies, 2017

Enterprise Immune System

Powered by machine learning and AI algorithms, the Enterprise Immune System developed by DarkTrace, Inc.[141] is advertised as the world's most advanced machine learning technology for cyber defence to date. In 2016, DarkTrace launched the first-ever autonomous response technology, Darktrace Antigena. It allowed the Enterprise Immune System to react to in-progress cyber-attacks in a highly precise way, giving security teams the time they desperately need to catch up. When the WannaCry ransomware attacks hit hundreds of organizations in 2017, Darktrace Antigena reacted in seconds, protecting customer networks from the inestimable damage.

In November 2017, the company announced a new business unit, Darktrace Industrial, dedicated to fighting threats in industrial and SCADA networks to protect critical national infrastructure and operational technology.

---

[141] DarkTrace, Cambridge, last retrieved 24 September 2018 at https://www.darktrace.com/technology/

The forecasts predict that the dual use of the AI-enhanced cyber technologies will keep affecting their application for both defence and offence, and their adaptivity and ability to act autonomously will only increase the speed and efficiency of attacks. This threat is discussed globally (e.g., the US Congressional hearings), and risk scenarios are analyzed (e.g., the subversion of military lethal autonomous weapon systems) to better guide research and develop response strategies.

IBM Research developed DeepLocker, "an ultra-targeted and evasive malware, to better understand how several existing AI models can be combined with current malware techniques to create a particularly challenging new breed of malware. This class of AI-powered evasive malware learns the rules, evade them and conceals its intent until it reaches a specific victim. It unleashes its malicious action as soon as the AI model identifies the target through indicators such as facial recognition, geolocation and voice recognition".[142]

DeepLocker: Ultra-Targeted and Evasive Malware

The use of AI makes malware almost impossible to reverse engineer. The malicious payload will only be unlocked if the intended target is reached. It achieves this by using a deep neural network (DNN) AI model.

The AI model is trained to behave normally unless it is presented with a specific input: the trigger conditions identifying specific victims. The neural network produces the "key" needed to unlock the attack. DeepLocker can leverage several attributes to identify its target, including visual, audio, geolocation and system-level features. When attackers attempt to infiltrate a target with malware, a stealthy, targeted attack needs to conceal two main components: the trigger condition(s) and the attack payload.

DeepLocker is able to leverage the "black-box" nature of the DNN AI model to conceal the trigger condition. A simple "if this, then that" trigger condition is transformed into a deep convolutional neural network of the AI model that is virtually impossible to decipher. In addition to that, it is able to convert the concealed trigger condition itself into a "password" or "key" that is required to unlock the attack payload.

Technically, this method allows three layers of attack concealment. That is, given a DeepLocker AI model alone, it is extremely difficult for malware analysts to figure out what class of target it is looking for.

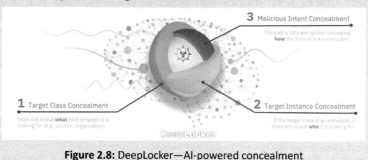

Figure 2.8: DeepLocker—AI-powered concealment

---

[142] M.Ph. Stoechlin, DeepLocker: How AI Can Power a Stealthy New Breed of Malware, 8 August 2018, last retrieved 17 August 2017 at https://securityintelligence.com/deeplocker-how-ai-can-power-a-stealthy-new-breed-of-malware/

> DeepLocker describes an entirely new class of malware—any number of AI models could be plugged in to find the intended victim, and different types of malware could be used as the "payload" that is hidden within the application.[143]
>
> AI-powered threats like DeepLocker are coming our way very soon;
>
> Attackers have the capability to build stealthy malware that can circumvent defenses commonly deployed today and;
>
> While a class of malware like DeepLocker has not been seen in the wild to date, these AI tools are publicly available, as are the malware techniques being employed—so it is only a matter of time before these tools combined by adversarial actors and cybercriminals become mainstream.
>
> *SecurityIntelligence*

As AI development is a very young field of studies, there is a lot of uncertainty in the issue of its potential, limits and control. Thus, AI might exhibit the unexpected behaviour because of the way it learns from large amounts of data is not entirely understood yet. That makes AI vulnerable to manipulation. Today's computer vision algorithms, for example, can be fooled and enabled to see and report things that do not correspond to reality (e.g., AI pixel poisoning).

The authors of the recently published Report on The Malicious Use of Artificial Intelligence: Forecasting, Prevention and Mitigation[144], provided a comprehensive illustration of the upcoming AI threats, in particular for:

- *Digital security*: The use of AI to automate tasks involved in carrying out cyberattacks will alleviate the existing trade-off between the scale and efficacy of attacks. This may expand the threat associated with labour-intensive cyberattacks (such as spear phishing). We also expect novel attacks that exploit human vulnerabilities (e.g. through the use of speech synthesis for impersonation), existing software vulnerabilities (e.g. through automated hacking), or the vulnerabilities of AI systems (e.g. through adversarial examples and data poisoning).
- *Physical security*: The use of AI to automate tasks involved in carrying out attacks with drones and other physical systems (e.g. through the deployment of autonomous weapons systems) may expand the threats associated with these attacks. We also expect novel attacks that subvert cyber-physical systems (e.g. causing autonomous vehicles to crash) or involve physical systems that it would be infeasible to direct remotely (e.g. a swarm of thousands of micro-drones).

---

[143] Marc Ph. Stoecklin, DeepLocker: How AI Can Power a Stealthy New Breed of Malware, 8 August 2018, last retrieved on 20 August 2018 at https://securityintelligence.com/deeplocker-how-ai-can-power-a-stealthy-new-breed-of-malware/

[144] Report on The Malicious Use of Artificial Intelligence: Forecasting, Prevention and Mitigation, February 2018, last retrieved on 21 August 2018 at http://img1.wsimg.com/blobby/go/3d82daa4-97fe-4096-9c6b-376b92c619de/downloads/1c6q2kc4v_50335.pdf

- *Political security*: The use of AI to automate tasks involved in surveillance (e.g. analysing mass-collected data), persuasion (e.g. creating targeted propaganda), and deception (e.g. manipulating videos) may expand threats associated with privacy invasion and social manipulation. We also expect novel attacks that take advantage of an improved capacity to analyse human behaviours, moods, and beliefs on the basis of available data. These concerns are most significant in the context of authoritarian states, but may also undermine the ability of democracies to sustain truthful public debates.

According to predictions, between 2016 and 2025, businesses will spend almost 2.5 billion USD on artificial intelligence to prevent cyberattacks.[145]

The AI-enhanced systems will keep helping humanity to solve many problems in their private and professional life. Their potential is unlimited and, hopefully, with prediction, modelling and the proactive prevention of all potential threats and complications, they will serve only to the benefits of humanity.

2.6.3 Game Theory Modelling

"Game Theory" is a term which is currently used to study logical decision-making in humans, animals and computer networks. It is an interdisciplinary field which has found application in many areas of modelling real life choices and behavioural relations. It is mainly used in economics, finance, political science, biology, and psychology; it is also applied in computer sciences, including cyber war modelling. Its concept is based on conflict and cooperation mathematical models[146], and is successfully used in security and cyber defence exercises.

It is widely considered that game theory as a mathematical discipline was founded in 1928 by Von Neumann with the proof of his minimax theorem on the mixed-strategy equilibria in two-person zero-sum games with perfect information. The theorem was called minimax as it showed that in this type of game there exists a pair of strategies for both players that allows each to minimize their maximum losses. Among all possible strategies each player should select the one optimal strategy that will result in the minimization of their maximum loss. His book "Theory of Games and Economic Behavior", co-authored by Oskar Morgenstern, enhanced the theorem and extended the theory for games with imperfect information and more than two players. This theory, based on the analysis

---

[145] 5 Cybersecurity Challenges and Trends: What to Expect in 2018, 10 January 2018, last retrieved on 17 August 2018 at https://www.globalsign.com/en/blog/cybersecurity-trends-and-challenges-2018/

[146] Roger B. Myerson Game Theory, Analysis of Conflict, 1991, Harvard University Press, USA.

of interrelationships of various coalitions which can be formed by the players of the game[147], grounded the foundation for treating decision-making under uncertainty.

Another fundamental concept of the Game Theory is a Nash equilibrium point, which in contrary to the above theory, is based on the absence of coalitions and assumes that each participant acts independently. This method is now one of the most widely used to predict the outcome of a strategic interaction of decision makers with a classic game structure which includes players, actions/strategies, and payoff for each player. Each strategy in the Nash equilibrium is the best response to all other strategies in that equilibrium, which is achieved when each player has identified a strategy, and no player can profit through the change of strategies while the other players keep their strategies unchanged. In 1951, in his article on "Non-Cooperative Games"[148] Nash defined a mixed strategy and proved that a finite non-cooperative game always has at least one equilibrium point. To predict the effectiveness, the choices of decision makers should be analyzed not in isolation but with consideration of the decisions of the others involved.

Game Theory allows us to analyse cyber threats, and has been used to provide hypothetical predictions and decision support in cyber security for a number of years. The approaches are numerous, e.g., a hybrid approach based on the use of the game theory and classical optimization for enhancing cyber security measures through producing decision support tools for ICT personnel.[149]

Cyber game modelling for mixed participation of human players and machines develops strategic decision making and is based on dynamic threat scenarios with proportional rewards. Game theory allows us to apply different techniques to implement the tactical analysis of threats, evaluate decisions through predicted outcomes, assess the vulnerabilities of attackers and test the professional skills of victims. It teaches us to find the best response options, including attribution capabilities—when to tolerate and when to counterattack. Computer capacities enhance the processes and allow us to explore all what-if combinations possible.

The beginning of the XXI century saw the explosion of cyber scenarios varying from simple incidents and DDoS attacks[150], with the analysis of

---

[147] J. Nash, Non-Cooperative Games, Annals of Mathematics, Second Series, Vol. 54, No. 2, (Sep., 1951), pp. 286-295, Published by: Mathematics Department, Princeton University DOI: 10.2307/1969529, last retrieved on 2 July 2018 at https://www.jstor.org/stable/1969529

[148] Ibid

[149] Ch. Hankin, Game Theory and Industrial Control Systems, Springer-Verlag New York, Inc. New York, NY, USA, 2016, doi>10.1007/978-3-319-27810-0_9 ISBN: 978-3-319-27809-4, last retrieved on 2 July 2018 at https://dl.acm.org/citation.cfm?id=2947452

[150] J. Mirkovic and P. Reiher, "A taxonomy of DDoS attack and DDoS defense mechanisms," SIGCOMM Comput. Commun. Rev., vol. 34, no. 2, pp. 39–53, 2004.

computer networks security[151], to information warfare[152], attacks on infrastructure.[153] The game theoretic model gained popularity through proposing the most efficient techniques to compute equilibria in games.[154]

With artificial intelligence coming on in the arena, game theory is applied in deep learning. Its concepts are used in the design of various deep learning architectures which are further tested in situations with imperfect knowledge, e.g. the DeepMind's AlphaGo. In the imperfect game scenario, the machine is supposed to learn from the provided data, improve itself and develop best strategies to reach the Nash Equilibrium without being informed about the choices of other players. In spite of a very complicated way of learning which is not very well understood yet, and the necessity to provide massive data sets it is a huge step towards the creation of artificial intelligence.

There are still multiple challenges in developing the game theoretical approach due to the multilayer complexity of cyber reality.

2.6.4 Cyber Security Trends

With the exponential growth of computer technologies, it is important to track the trends to be better prepared to respond to emerging challenges.

Highlights

AI-powered Attacks

AI-based software can "learn" from the knowledge bases and outcomes of the past events, thus being able to predict and identify cybersecurity threats. It can also develop potential attacks scenarios. These capacities may be used by malicious actors to launch even more sophisticated cyber-attacks in the future.

Sandbox-evading Malware

Researchers and companies increasingly use sandboxing and virtualization technology for detecting and preventing malware infections. However, cyber-criminals are finding more ways to evade this technology. For example, new strains of malware are able to recognise when they are inside a sandbox or any virtual container, and do not trigger the malicious code until they are outside of the sandbox.

---

[151] K. Lye and J. Wing, "Game Strategies in Network Security," Copenhagen, Denmark, 2002.

[152] S.N. Hamilton, W.L. Miller, A. Ott and O.S. Saydjari, Challenges in Applying Game Theory to the Domain of Information Warfare, Proc. 4th Inf. Surviv. Work., 2002.

[153] A. Chakrabarti and G. Manimaran, "Internet Infrastructure Security: A Taxonomy," IEEE Netw., no. December, pp. 13–21, 2002

[154] A. Fielder, E. Panaousis, P. Malacaria, C. Hankin, and F. Smeraldi, "Game Theory Meets Information Security Management," in Information Security and Privacy Conference, 2014, 2014, pp. 15–29. Quoted from A. Chukwudi et al., Game Theory Basics and Its Application in Cyber Security, 2017, last retrieved on 20 August 2018 at http://article.sciencepublishinggroup.com/pdf/10.11648.j.awcn.20170304.13.pdf

Ransomware and Internet of Things

Most IoT devices do not typically store valuable data and the ransomware designed for those devices is less popular. Developing ransomware for IoT devices would not be cost effective as the potential number of victims would be lower than from the ransomware for personal computers or smartphones.

However, with growing IoT application, the potential damage from the IoT ransomware could be much higher in the near future. For example, hackers may choose to target valuable systems, such as a building's climate control system in an extremely hot or cold climate geographical zone, big food refrigeration units, wearable devices, security alarm systems, etc.

Multi-factor Authentication

According to the 2016 Data Breach Investigations Report by Verizon[155], "63% of confirmed data breaches involved leveraging weak, stolen or default passwords." This is largely due to the fact that most organizations are still using single-factor authentication, which basically relies solely on a single password or PIN.

Companies tend to postpone the implementation of multi-factor authentication (MFA), as it would negatively affect user experience. However, according to the research carried out by Bitdefender[156], there is a growing concern about stolen identities amongst the general public. As such, there will likely be an increase in the number of companies implementing MFA.

More Sophisticated Security Technologies

The use of "remote browsers" can be helpful for isolating a user's browsing session from the network/endpoints.

More advanced virtualization and deception technologies will function by imitating a company's critical assets using honeypot software and will act as a trap for attackers.

There will also be an increase in the use of solutions which can detect and respond to anomalous behaviour using AI and Machine Learning-based intrusion detection systems.

A Rise of State-sponsored Attacks

Cyber threats keep growing in complexity and cyber arms in availability.

---

[155] 2016 Data Breach Investigations Report, Verizon, 2016, last retrieved on 5 September 2018 at https://www.verizonenterprise.com/resources/reports/rp_DBIR_2016_Report_en_xg.pdf

[156] Bitdefender Total Security 2018: User's Guide, Bitdefender, 18 August 2018, last retrieved on 5 September 2018 at https://www.bitdefender.co.th/resources/home-user/bitdefender_ts_uniguide_2018_en.pdf

The rise of state sponsored cyberattacks is one of the most concerning areas of cyber-security. Their goals vary—from intelligence gathering to sabotage, and may further target electronic voting systems to manipulate public opinion, critical infrastructure, military command and control systems to facilitate conventional warfare.

Automated Audit

To test the network for potential publicly known vulnerabilities, organizations prefer using automated auditing technology and apply vulnerability scanning software (e.g., Nassus, OpenVAS). In addition, configuration analysis software can be used to identify potential configuration issues and related security flaws.

However, the use of the public vulnerability scanners might not produce any results in the military information technology infrastructure. Their vulnerability databases contain only publicly available vulnerabilities, while a military installation might use custom software which is not a part of the public domain.

In contrast, advanced configuration analysis tools scan individual network infrastructure components, use virtual modelling to find deep structural vulnerabilities, and generate systematic reports on how to improve defences. These technologies can allow entire military bases to be audited in minutes, a process that could take weeks for human auditors.

New Regulations

New bilateral and multilateral legal acts, norms, and standards will keep being introduced in the near future. For example, the new EU regulation General Data Protection Regulation (GDPR), which came into effect on 25 May 2018, offers a number of important changes to the current Data Protection Directive. These include increased territorial scope, stricter consent laws and elevated rights for data subjects to name a few.

The national and international enforcement of new cyber regulations will be enhanced.

2.7 Cyber Arms Race and Control

The art of war teaches us to rely not on the likelihood of the enemy's not coming, but on our own readiness to receive him;

Sun Tzu, Art of War, Book VIII

For about a century, arms races, characterized by rapid threat-driven and competitive development of military power, have been the subject of study of international relations, and findings evidence a close relation between arms races escalation and wars.

The Online Cambridge dictionary defines the arms race as a situation in which two or more countries are increasing the number and strength of their weapons.[157] It might be also defined as "the participation of two or more nation-states in apparently competitive or interactive increases in quantity or quality of war material and/or persons under arms."[158] The term is generally used to describe a competition where the goal is to stay ahead of the other states, either by intentionally escalating the strength to become the first and strongest and deter everybody around, or from fear of being overcome by the rival states and a sense of overall insecurity.

Today we are observing an on-going militarization of cyberspace and high escalation risks.[159] Yet until today, the issue of a global cyber arms race has not been comprehensively analyzed in the academic sense of the word. The following reasons, complicating the clear perception of the cyber arms race, can be named:

- The definition of "Cyber arms" is still not clear due to their dual character;
- Virtual cyber weapons - anonymous, invisible and easily disguisable as security tools - do not require any stocks;
- There are no cyber "armoury parades" demonstrating cyber capabilities;
- State budgets do not always provide budget lines for the specific development of cyber tools, and these are not always evident, considering that the price of accessible cyber tools goes down;
- Lack of cyberspace geographical constraints.

2.7.1 Evaluation Approach to Cyber Arms Race

Opinions on cyber threats vary, from highlighting the potential of global damage and the need to elevate the issue to the top of the state's national security[160,161], to criticizing the false inflation and disconnection from reality[162] both from fear and an exaggeration of stockpile. Here, it is worth noting that fear and premonitions can be just as powerful drivers

---

[157] Definition—arms race, Cambridge Dictionary, last retrieved on 15 October 2018 at https://dictionary.cambridge.org/dictionary/english/arms-race

[158] Smith, Theresa Clair (1980). "Arms Race Instability and War". Journal of Conflict Resolution. 24 (2): 253–284. doi:10.1177/002200278002400204

[159] M. Hansel, Cyber-attacks and psychological IR Perspectives: Explaining Misperceptions and Escalation Risks, Journal of International Relations and Development, 2017, doi:10.1057/s41268-016-0075-8

[160] Richard A. Clarke and Robert Knake, 2010. Cyber War: The Next Threat to National Security and What to Do About It.Harper Collins Publishers New York, NY.

[161] Lucas Kello, "The Meaning of the Cyber Revolution: Perils to Theory and Statecraft." Quarterly Journal: International Security, vol. 38. no. 2. (Fall 2013): 7-40

[162] B. Valeriano and R.C. Maness, 2015, Cyber War versus Cyber Realities, Oxford University Press, Oxford.

of security competition as actual threats. Media reports frequently use the term 'arms race' to describe the global proliferation of cyber warfare capabilities as states respond to their security concerns." [163]

It is commonly accepted that in the insecure cyberspace environment, unpredictable covert cyberattacks, push the states to escalate the cyber arms race, invest into cyber research and development, and gather cyber intelligence. Unlike the physical warfare domain, the virtual nature of malware makes it very difficult for states to gain an accurate picture of one another's capabilities. The anonymity that cyber methods can provide the attacker and the resulting attribution problems add to this uncertainty.

The nature of cyberspace means that cyber weapons are much easier to steal, smuggle or replicate than conventional armaments. As it is easier and less costly to copy lines of code than it is to steal a physical weapon, criminals increasingly collect arsenals of cyber arms and sell them in the digital black markets. Within those well-developed trading platforms, criminals barter the cyber equivalent of smart bombs and nuclear devices, many of which have user guides, money-back guarantees and user reviews with ratings. Powerful custom-made tools designed to exploit unpatched (and usually unknown) vulnerabilities and autonomously reproduce themselves across the world are widely available to buy, rent or franchise. These tools also can be repurposed and customized for a particular task, from hacking a warship[164] to stealing personal health data from a hospital.

Craig and Valeriano in a paper on "Conceptualizing cyber arms races" investigated in detail the emergence of an arms race dynamic in the international cyber domain and proposed methods for data collection. They identified the accounting methodology for the build-up of cyber capabilities by nation states, examined the concept of the cyber arms race, portrayed a state behaviour in the cyber domain and provided an example of a macro study by examining the cases of the United States and Iran, and of North Korea and South Korea, which are among the major players in the cyber conflict arena. Time series data was employed on a number of indicators to measure each state's scale of increase in cyber capabilities and the findings suggested that these state dyads have indeed been engaged in cyber arms races, as defined by their competitive and above-normal mutual increase in cyber capabilities.[165]

This proposed methodology could be very helpful in establishing a pathway for the future extensive data collection of this new phenomenon.

---

[163] McAfee Threats Report: Second Quarter 2012, last retrieved on 5 September 2018 at https://www.cise.ufl.edu/~nemo/cybersecurity/rp-quarterly-threat-q2-2012.pdf

[164] S. Abaimov and P. Ingram, Hacking UK Trident: The Growing Threat, 2017, last retrieved 2 September 2018 at http://www.basicint.org/sites/default/files/HACKING_UK_TRIDENT.pdf

[165] A. Craig and B. Valeriano, Conceptualizing cyber arms race, 2016 8th International Conference on Cyber Conflict, 2016, last retrieved 2 September 2018 at http://www.ccdcoe.org/cycon/2016/proceedings/10_craig_valeriano.pdf

But the authors finally conclude that it is not possible to quantify the actual cyber 'arms' or malware possessed by states, and they acknowledged this limitation. Also, stating that the current lack of data in this relatively new and often secretive domain means that alternative methods for evaluating the magnitude of cyber build-ups will need to be used.[166]

The traditional approach of comparing the cyber arms race based on only two states might be very misleading in relation to the global character of cyberspace coverage, covert cyber space channels without any "country" borders, disguised IP addresses, etc. This issue should be considered globally, starting with the analysis of expenditures and cyber dynamics by major cyber powers, such as China, Israel, Russia, United States, United Kingdom, as well as major international cooperation programmes, etc.

When evaluating the states' intentions, closer attention should be paid to the following:

1. Overall routine government spending on cyber security
2. Number of cyber security specialists employed by governments
3. Research and development of:
   - software
   - autonomous devices and robots
   - augmentations and enhancements for human soldiers
   - state-of-the-art technologies, including AI, and machine learning
4. Cyber security tests (timing, funding, frameworks)
5. Cyber technologies application
6. Cooperation with big commercial and multinational entities
7. Cyber benefits from new technology
8. International coalitions for cyber development and their funding
9. Research and collaboration programmes

Since cyber technologies can be much cheaper than conventional weapons, weaker states can gain asymmetric advantages by entering into the cyber arms arena and compete on a more even ground with traditionally powerful states. The sources of threat are therefore potentially more widespread. Insecurity is also supported by the belief that the cyber conflict domain favours the offense.

In the dilemma of defensive versus offensive, the offensive approach wins. The offense-defence balance theory postulates that if offensive military capabilities hold advantages over defensive capabilities, the security dilemma is more intense and the risk of an arms race and war is greater. Offensive cyber capabilities are assumed to be more cost effective and efficient, whereas defence is difficult given the immense challenge involved in securing every civilian and privately-owned critical

---

[166] Ibid

network and closing every vulnerability possible, since many of which go undetected until an attack reveals them.

The Internet's lack of geographical constraints further decreases its defence effectiveness, as any cyber conflict may immediately escalate to global levels. Thus, offensive preparations may therefore become the dominant strategy of any state.

In February, 2018, Nicola Whiting, the chief operating officer of Titania Group, voted one of the top 20 most influential women (by SC Magazine) working in cybersecurity in the United Kingdom, wrote: "Advances in automated cyber weapons are fuelling the fires of war in cyberspace and enabling criminals and malicious nation-states to launch devastating attacks against thinly stretched human defences. Allied forces must collaborate and deploy best-of-breed evaluation, validation and remediation technologies just to remain even in an escalating cyber arms race".[167]
She further states, "the wholesale investment in and propagation of cyber weaponry is behind the growing scale and severity of the threat forces face today. An arms race for offensive cyber capabilities among government, terrorist and other groups has resulted in a digital cold war with a goal of global dominance that has, up until now, only been achievable with conventional weapons".

In analysing the cyber arms race, due attention should be paid to the funding of "civil" programmes for technologies development. Thus, with this in mind recently, the European Commission proposed to create the first ever Digital Europe programme and invest 9.2 billion euros to align the next long-term European Union budget 2021–2027 with increasing digital challenges.[168] It is meant at increasing the EU's international competitiveness as well as developing and reinforcing Europe's strategic digital capacities. These key capacities concern high-performance computing, artificial intelligence, cybersecurity and advanced digital skills and ensuring their wide use and accessibility across the economy and society by businesses and the public sector alike.

Considering the current trends, we may assume that the cyber arms race will be escalating further. New and more sophisticated intuitive cyber tools will appear on the arena, reinforced by AI advantages. At the same time, it will boost the development of defensive technologies, which at a certain stage of AI might start replacing human operators.

In 2017 (ISC)2 Global Information Security Workforce Study[169], considered to be one of the largest surveys of the global cybersecurity

---

[167] N. Whiting, Cyberspace Triggers a New Kind of Arms Race, 1 February 2018, last retrieved on 15 July 2018 at https://www.afcea.org/content/cyberspace-triggers-new-kind-arms-race

[168] EU budget: Commission proposes €9.2 billion investment in first ever digital programme, Brussels, 6 June 2018, last retrieved on 16 July 2017 at http://europa.eu/rapid/press-release_IP-18-4043_en.htm

[169] Booz, Allen, Hamilton, 2017 Global Information Security Workforce Study, A Frost & Sullivan Executive Briefing, June 2017, last retrieved on 15 July 2018 at https://iamcybersafe.org/wp-content/uploads/2017/06/Europe-GISWS-Report.pdf

work force, predicted a shortfall of 1.8 million cyber security workers by 2022. The increasing use of automated cyber warfare weapons, coupled with the shortage of defenders, means military armed forces are fighting an uphill battle to protect themselves online.

The 2018 World Economic Forum Global Risks Report warned against complacency and reminded us that risks can crystallize with disorientating speed. In a world of complex and interconnected systems, feedback loops, threshold effects and cascading disruptions can lead to sudden and dramatic breakdowns.[170]

Further exploration of the concept of the cyber arms race is a vital necessity considering the increasing attention to the development of AI, machine learning, robotics and merge of cyber tools with Weapons of Mass Destruction (WMD). Additionally, AI could potentially be beneficial to model the states' reaction to cyber security threats and predict the future technological directions for simulation exercises and further policy development.

2.7.2 Artificial Intelligence Arms Race

The next circle of the cyber arms race is classified as the AI arms race. In theory, the only technology capable of hacking a system run by artificial intelligence is another, more powerful AI system.

> "It's an arms race," said Walter O'Brien, CEO of Scorpion Computer Services, whose AI system runs and protects the Army's UAV operations. "Now I have an AI protecting the data centre, and now the enemy would have to have an AI to attack my AI, and now it's which AI is smarter."[171]

Coined as the next space race, the race for AI dominance is both intense and necessary for nations to remain in primary position in an evolving global environment. As technology develops the amount of virtual information and the ability to operate at optimal levels when taking advantage of this data also increases. Furthermore, the proper use and implementation of AI can facilitate a nation in the achievement of information, economic, and military superiority—all ingredients towards maintaining a prominent place on the global stage.[172]

It is not easy to measure the AI capabilities of countries. The tools, technologies and know-how are all "dual-use"—they lie scattered

---

[170] The Global Risks Report 2018, 13th edition, World Economic Forum, 2018, last retrieved on 15 July 2018 at http://www3.weforum.org/docs/WEF_GRR18_Report.pdf

[171] M.R. Bauer, The next cyber arms race is an artificial intelligence, 24 January 2018, last retrieved on 16 July 2018 at https://www.fifthdomain.com/dod/2018/01/24/the-next-cyber-arms-race-is-in-artificial-intelligence/

[172] E. Garcia, The Artificial Intelligence Race: U.S. China and Russia, Modern Diplomacy, 19 April 2018, last retrieved on 16 July 2018 at https://moderndiplomacy.eu/2018/04/19/the-artificial-intelligence-race-u-s-china-and-russia/

across civilian and military spheres, in locations around the world. Understanding a country's relative AI power requires a deep knowledge of both the public and private sectors, with information often classified or deliberately misleading.[173]

Due to the considerable advantages artificial intelligence can provide, there is now a race between the major powers and coalitions to master AI and integrate this capability into military applications in order to assert power and influence globally. China publicly set the goal of becoming "the world's primary AI innovation centre" by 2030. The President of Russia said that "artificial intelligence is the future, not only for Russia but for all humankind... Whoever becomes the leader in this sphere will become the ruler of the world".[174]

Russia

In 2017, publicly announced AI investments in Russia equalled 700 million roubles (12.5 million USD) with the prognosis to grow up to 28 billion roubles (500 million USD) by 2020.[175]

Though the officially published figures are not high, Russia is actively competing in the AI arms race by developing artificially intelligent missiles, drones, unmanned vehicles, military robots and medic robots.

In 2017, General Viktor Bondarev, commander-in-chief of the Russian air force, revealed to the press that Russia has been working on AI-guided missiles with a range of up to 7,000 kilometres which can analyse the aerial and radio-radar situation and determine its direction, altitude and speed.[176]

The same year, the CEO of Russia's Kronstadt Group, a defence contractor, told Russia's state news agency TASS that "there already exist completely autonomous AI operation systems that provide the means for UAV clusters, when they fulfil missions autonomously, sharing tasks between them, and interact", and added that his company is also working on a drone defence system, something he said will become "obligatory."[177]

---

[173] T. Upchurch, How China could beat the West in the deadly race for AI weapons, 8 August 2018, last retrieved on 17 August 2018 at https://www.wired.co.uk/article/artificial-intelligence-weapons-warfare-project-maven-google-china

[174] R. Cardillo, Racing to secure our future, The Cipher Brief, 22 May 2018, last retrieved on 15 July 2018 at https://www.thecipherbrief.com/column/agenda-setter/racing-secure-future

[175] Рынок искусственного интеллекта в России оценили в 700 миллионов/Russia's AI market is worth 700 millions, 27 November 2017, last retrieved on 15 July 2018 at http://www.cnews.ru/news/top/2017-11-27_rynok_iskusstvennogo_intellekta_v_rossii_otsenivaetsya

[176] Tom O'Connor, Russia's military challenges U.S. and China by building a missile that makes its own decisions, 20 July 2017, last retrieved on 18 November 2018 at https://www.newsweek.com/russia-military-challenge-us-china-missile-own-decisions-639926

[177] Jason Le Miere, Russia developing autonomous 'swarm of drones' in new arms race with U.S., China, 15 May 2017, last retrieved on 7 October 2018 at https://www.newsweek.com/drones-swarm-autonomous-russia-robots-609399

In 2018, the US Nuclear Posture Review[178] alleged that Russia is developing a "new intercontinental, nuclear-armed, nuclear-powered, undersea autonomous torpedo" named "Status 6".

The AI-related events show the increased government attention to the area. In March 2018, in the opening speech to the first AI conference "Artificial Intelligence: challenges and solution", the Defence Minister Sergei Shoigu urged military and civilian scientists to join forces in developing AI-technologies to counter possible threats in the fields of technological and economic security in Russia.[179]

The conference was a milestone in the open discussion of the role of AI in the defence military application. The Deputy Defence Minister, Yury Borisov stated: "Today cyberwar is a reality. Current battles are not on the battlefields, they are initially staged in the information space. The Winner is the party which is able to control it, to counterattack in the proper way". The representative of the Russian Academy of Sciences highlighted that the AI technologies should target the speed and quality of military solutions, such as the field assessment, decision making, counter activities in information space, military defence and offence operations.[180]

Many of the developed AI-related projects are under the auspices of the Ministry of Defense (MOD) and its affiliated institutions and research centres. A three-year-old AI and semantic data analysis research project led by the military-related United Instrument-Making Corporation—and involving the Russian Academy of Sciences, various universities, and more than 30 private companies—is shaping up to be one of Russia's biggest public-private approaches. Earlier this year, MOD announced a competition amongst the "designers of robotics technologies" to develop artificial intelligence. In March, Russia's six-year-old Foundation for Advanced Studies—created as a parallel to the Pentagon's Defense Advanced Research Project Agency—announced proposals for MOD to standardize artificial intelligence development along four lines of effort: image recognition, speech recognition, control of autonomous military systems, and information support for weapons' life-cycle.[181]

---

[178] Nuclear Posture Review, Office of the Secretary of Defense, February 2018, last retrieved on 2 September 2018 at https://media.defense.gov/2018/Feb/02/2001872886/-1/-1/1/2018-NUCLEAR-POSTURE-REVIEW-FINAL-REPORT.PDF

[179] Шойгу призвал военных и гражданских ученых совместно разрабатывать роботов и беспилотники/Shoigu urged military and civilian academics to jointly develop robots and unmanned vehicles, 14 March 2018, TASS, last retrieved on 7 December 2018 at https://tass.ru/armiya-i-opk/5028777

[180] Outcomes of the conference "Artificial intelligence, challenges and solutions", 14-15 March 2018, published on 22 March 2018, last retrieved on 8 December 2018 at https://ya-r.ru/2018/03/22/itogi-konferentsii-iskusstvennyj-intellekt-problemy-i-puti-resheniya-14-15-marta-v-parke-patriot/

[181] S. Bendett, In AI, Russia is hustling to catch up, Defense One, 4 April 2018, last retrieved on 8 December 2018 at https://www.defenseone.com/ideas/2018/04/russia-races-forward-ai-development/147178/

China

China is already considered to be the world's second-largest investor in artificial intelligence, and has set out an ambitious plan to surpass the United States in being the global leader in this field by 2030.[182]

China's State Council published a document in July of 2017 entitled, "New Generation Artificial Intelligence Development Plan." It laid out a plan that takes a top-down approach to explicitly map out the nation's development of AI, including goals reaching all the way to 2030. It takes a three-step approach: firstly, it keeps pace with leading AI technology and applications in general by 2020, in order to make major breakthroughs by 2025, and be the world leader in the field five years thereafter. Up to 26 per cent of China's GDP could be generated by AI-related industries by 2030, making the country the world's biggest winner from investing in this field, according to a report by PricewaterhouseCoopers.[183]

United States

In 2014, the former Secretary of Defense Chuck Hagel posited the "Third Offset Strategy" that rapid advances in AI will define the next generation of warfare[184]. According to data science and analytics firm Govini, the US Department of Defense increased investment in artificial intelligence, big data and cloud computing from 5.6 billion USD in 2011 to 7.4 billion USD in 2016.

The US is still considered to be at the forefront of AI research, leading the way in industrial and military applications of AI. The country retains dominance across many of the standard metrics used as proxies to evaluate AI power, particularly intellectual talent, research breakthroughs and superior hardware.[185]

Japan Times reported[186] in 2018 that the US's private investment is around 70 billion USD per year. At the same time, governments are investing in foundational research to expand the scope in capabilities of AI systems. In 2016, DARPA hosted the Cyber Grand Challenge contest,

[182] A. Lee, World dominance in three steps: China sets out road map to lead in artificial intelligence by 2030, 21 July 2017, last retrieved on 8 July 2018 at https://www.scmp.com/tech/enterprises/article/2103568world-dominance-three-steps-china-sets-out-road-map-lead-artificial

[183] Ibid

[184] Z. Cohen, US risks losing artificial intelligence arms race to China and Russia, 29 November 2017, last retrieved on 19 October 2018 at www.cnn.com/2017/11/29/politics/us-military-artificial-intelligence-russia-china/index.html

[185] T. Upchurch, How China could beat the West in the deadly race for AI weapons, 8 August 2018, last retrieved on 30 August 2018 at https://www.wired.co.uk/article/artificial-intelligence-weapons-warfare-project-maven-google-china

[186] The artificial intelligence race heats up, 1 March 2018, The Japan Times, last retrieved on 30 August 2018 at https://www.japantimes.co.jp/opinion/2018/03/01/editorials/artificial-intelligence-race-heats/

which saw teams of human researchers compete with each other to create programs that could autonomously attack other systems while defending themselves.

> During the April 9th Washington, DC conference on the future of war sponsored by the non-profit think tank New America, M. Griffin, the Undersecretary of Defense for research and engineering, called for more serious work by the Pentagon, saying, "There might be an artificial intelligence arms race, but we are not yet in it." America's adversaries, he said, "understand very well the possible utility of machine learning. I think it's time we did as well."[187]

The continuous development and improvement of AI requires a comprehensive plan and partnership with industry and academia. To target this issue two DOD-directed studies, the Defense Science Board Summer Study on Autonomy and the Long-Range Research and Development Planning Program, highlighted five critical areas for improvement:[188]

- autonomous deep-learning systems,
- human-machine collaboration,
- assisted human operations,
- advanced human-machine combat teaming,
- network-enabled semi-autonomous weapons.

France

In March 2018, French President Emmanuel Macron promised 1.5 billion euros (1.85 billion USD) of public funding into artificial intelligence by 2022 in a bid to reverse a "brain drain" and catch up with the dominant US and Chinese tech giants.[189]

The President expressed his wish to turn France into a "startup nation" and bets that easing labour laws and higher investment technology will create jobs, alleviate the domination of Alphabet's Google, Facebook and lay out the seeds for Europe-based champions.

"There's no chance of controlling any effects (of these technologies) or having a say on any adverse effect if we've missed the start of the war," the president said in front of ministers and top executives.[190]

---

[187] Matt Stroud, The Pentagon is getting serious about AI weapons, 12 April 2018, last retrieved on 19 October 2018, https://www.theverge.com/2018/4/12/17229150/pentagon-project-maven-ai-google-war-military

[188] The Artificial Intelligence Race: US, China, Russia, by Ecatarina Garcia, Modern Diplomacy, April 19, 2018, last retrieved on 23 October 2018, https://moderndiplomacy.eu/2018/04/19/the-artificial-intelligence-race-u-s-china-and-russia/

[189] France to spend $1.8 billion on AI to compete with US, China, 29 March 2018, last retrieved on 19 October 2018, https://www.reuters.com/article/us-france-tech/france-to-spend-1-8-billion-on-ai-to-compete-with-u-s-china-idUSKBN1H51XP

[190] Ibid

President Macron's announced AI plan calls for the opening up of data collected by state-owned organizations such as France's centralized healthcare system to drive up efficiency through AI—the field in computer science that aims to create machines able to perceive the environment and make decisions.

United Kingdom

The United Kingdom is planning to invest in artificial intelligence technologies nearly 1 billion GBP (1.3 billion USD). The part of a multi-year AI investment—about 300 million GBP (more than 400 million USD) would come from UK-based corporations and investment firms and those located outside the country.[191] Some of the US-based companies involved with the UK's AI agreement include Microsoft, Hewlett Packard Enterprise, IBM, McKinsey, and Pfizer.

As part of the deal, the Japanese firm Global Brain plans to invest about 48 million USD in UK tech start-ups and will open a European headquarters in the United Kingdom. The University of Cambridge will also give UK businesses access to a new 13 million USD supercomputer to help with AI-related projects.

Canadian venture capital firm Chrysalix will also open a European headquarters in the UK and plans to invest more than 100 million USD in local start-ups specializing in AI and robotics.

NATO

Cyber Hunters - Autonomous and intelligent cyber defence agents[192]

In a conflict with a technically sophisticated adversary, NATO military tactical networks will operate in a heavily contested battlefield. Enemy software cyber agents—malware—will infiltrate friendly networks and attack friendly command, control, communications, computers, intelligence, surveillance, and reconnaissance (C4ISR) and computerized weapon systems. To fight them, NATO needs artificial cyber hunters—intelligent, autonomous, mobile agents specialized in active cyber defense.

In 2016, NATO initiated RTG IST-152 "Intelligent Autonomous Agents for Cyber Defense and Resilience".

Its objective is to help accelerate development and transition to the practice of such software agents by producing a reference architecture and technical roadmap.

If such research is successful, it will lead to technologies that enable the following vision. NATO agents will stealthily patrol the networks, detect the enemy agents while remaining concealed, and then destroy or degrade the enemy malware. They will do

---

[191] J. Vanian, United Kingdom Plans $1.3 Billion Artificial Intelligence Push, Fortune, 25 April 2018, last retrieved on 19 October 2018 at http://fortune.com/2018/04/25/uk-ai-artificial-intelligence-deal/

[192] Alexander Kott et al., Initial Reference Architecture of an Intelligent Autonomous Agent for Cyber Defence, NATO Report, ARL-TR-8337, US Army Research Laboratory, March 2018, last retrieved on 10 September 2018 at https://arxiv.org/ftp/arxiv/papers/1803/1803.10664.pdf

so mostly autonomously, because human cyber experts will be always scarce on the battlefield. They will be adaptive because enemy malware is constantly evolving. They will be stealthy because the enemy malware will try to find and kill them. At this time, such capabilities remain unavailable for the defensive purposes of NATO.

The IST-152 group is using a comprehensive, focused technical analysis to produce a first-ever reference architecture and technical roadmap for autonomous cyber defence agents. In addition, the RTG is working to identify and demonstrate selected elements of such capabilities, which are beginning to appear in academic and industrial research.

There is no concerted solution at the moment, whether to continue or ban the future development of AI for military purposes. On the one side are those that believe pursuing the development of military AI will lead to an unstoppable arms race and should be banned. Sandro Gaycken, a senior advisor to NATO, and a founder of the Digital Society Institute at ESMT, a Berlin-based business school, states that such initiatives are supremely complacent and risk granting authoritarian states an asymmetric advantage. Prohibiting AI research for military purposes will not lead to peace but give the upper hand to authoritarian systems. Therefore, if the West wants to stay in the lead, it needs to unify around a concerted strategy. "Within most military and intelligence organizations it's a concern," Gaycken argues. "And it's about to become a much larger concern."[193]

Humanity is still far away from developing artificial general intelligence, but the legal and ethical implications should be planned already, and considering measures for the responsible supervision, regulation, and governance of the design and deployment of AI systems. We need to heed the call by prominent AI researchers to commit to the careful monitoring and deliberation about the implications of AI advances for defence and warfare, including potentially destabilizing developments and deployments. Artificial super-intelligence—i.e. intelligence that is superior to that of homosapiens—is likely to pose quite unique challenges to defence and security planners.[194]

2.7.3 Cyber Arms Control Challenges

As weapons technology determines victory or defence in war, states have strong reasons to invest significant resources in the research on and

---

[193] Tom Upchurch, How China could beat the West in the deadly race for AI weapons, 8 August 2018, last retrieved on 30 August 2018, last retrieved on 19 October 2018 at https://www.wired.co.uk/article/artificial-intelligence-weapons-warfare-project-maven-google-china

[194] S. de Spiegeleire, M. Maas, T. Sweijis, Artificial Intelligence and the future of defense, The Hague Centre for Strategic Studies, 2017, last retrieved on 19 October 2018 at https://hcss.nl/sites/default/files/files/reports/Artificial%20Intelligence%20and%20the%20Future%20of%20Defense.pdf

development of cutting-edge weapon technology. Increased competition leads to an arms race and proliferation of new and dangerous weapon systems so that everyone's security is worse than it was before. This is the well-known logic of the security dilemma in its qualitative dimension.[195]

Since shortly after the first and, so far, only use of atomic weapons, in 1945, scientists, policy analysts, and government officials have sought to identify measures to inhibit the further acquisition and use of the destructive potential of nuclear technology. In the late 1960s, a similar group of stakeholder-initiated efforts to prevent the biological sciences from being used to develop weapons whose destructive effects against humans, animals, and plants could, in some circumstances, rival those of nuclear weapons. Today, questions are being raised about how to manage the potential threat posed by information technology, whose growth and spread some believe may position cyber weapons alongside nuclear and biological weapons in the elite club of technologies capable of unleashing massive harm.[196]

During the Cold War, when the nuclear-armed superpowers reached the level of mutually assured destruction threatening human existence, the arms controls efforts, especially related to the United States and the Soviet Union competition, turned out to be useful and lifesaving. Among the agreements there was the Partial Test Ban Treaty signed in 1963 between the US, the Soviet Union and the UK, which was later joined by more than one hundred countries; the Strategic Arms Limitation Talks Treaty (SALT I), signed in 1972 and consisting of the Anti-Ballistic Missile Treaty (ABM Treaty) and the Interim Agreement on the Limitation of Strategic Offensive Arms, further followed by SALT II, which limited nuclear arsenals. The 1990s saw the succession of the Strategic Arms Reduction Treaties on the limitation of the strategic offensive arms—START I, START II, START III, with the New START treaty signed in 2010 by the US and Russian Presidents to reduce strategic nuclear missile launchers by fifty percent and curtail the deployed nuclear warheads.

> Arms control, any international control or limitation of the development, testing, production, deployment, or use of weapons based on the premise that the continued existence of certain national military establishments is inevitable. The concept implies some form of collaboration between generally competitive or antagonistic states in areas of military policy to diminish the likelihood of war or, should war occur, to limit its destructiveness.[197]

---

[195] M. Hansel et al., Taming cyberwarfare: lessons from preventive arms control, Journal of Cyber Policy, Chatham House, Issue 1, Volume 3, 2018, ISSN 2573-8671, p. 44

[196] Governance of Dual-Use Technologies: Theory and Practice, Edited by Elisa D. Harris, American Academy of Arts and Sciences, Cambridge, 2016, last retrieved on 19 October 2018 at https://www.amacad.org/content/publications/pubContent.aspx?d=22234

[197] K.W. Thompson, Arms control, Encyclopaedia Britannica, last retrieved on 1 July 2018 at https://www.britannica.com/topic/arms-control

The transfer of the accumulated experience is proposed for replication in the cyber landscape. However, this model is comparatively clear with conventional weapons, and this will not necessarily be applicable for the reality of cyberspace. Opinions vary and currently there is no joint agreement on the issue, as there is no clear understanding on what exactly should be controlled and what could be controlled. The classification proposed at the beginning of the book seems to us very useful in tackling this challenge.

The proposed classification of cyber related instruments into cyber arms, cyber tools, cyber weapons and cyber-physical weapons provides a clear vision and comprehensive approach to the issue of cyber space control, with certain limitations though based on the assumptions that the absolutely clear definitions and instructions are not possible.

While cyber-physical weapons are feasible and, similar to conventional weapons, and can be enumerated, demonstrated, stockpiled, disassembled and replaced by upgraded versions; cyber-weapons can be recognized through malicious intentions only; cyber arms, being of dual use and consisting of exponentially evolving stealthy codes, which can be stolen, copied, regenerated, modified, are absolutely out of any imaginable strict control, or monitoring, unless just simply banned.

Thus, we see the major cyber arms control problem for the actual "cyber arms"—these "intelligent" pieces of malicious codes able to cause loss of strategic information and sabotage of strategic infrastructure. The challenges are the following:

- lack of clear definitions
- dual-use of cyber arms, and inefficiency of halting technological progress in non-military sectors
- stealthy and evolving nature of cyber arms
- difficulty in measuring the super powers capabilities in cyberspace
- agreements entail monitoring and control, which is challenging for cyber space
- agreements require enforcement, which is challenging in cyber space

The table below (Table 2.3) provides a comparative analysis of control in conventional and cyber arms.

As seen from the figure above, the application of conventional approaches to cyber arms control is next to impossible at the current stage of technological development. However, with the confidence building measures, openness and information sharing for research and development, global initiatives and bilateral agreements, all this is already a step forward in ensuring stability. Best practices in bilateral and multilateral agreement, development and tests of standards, terminology, regulations, certification, licencing, etc.—all of this contributes to the application of all the above on the international level.

Table 2.3: Comparative analysis of control in nuclear and cyber arms

Criteria	Nuclear arms	Cyber-physical arms	Cyber arms/ weapons
Quantitative criteria			
Measurement of strength	Possible, e.g., by number of missile warheads	Possible, e.g., weaponry, battery power, area of activities	N/A as cyber weapons can be continuously copied and regenerated
Lethality	Universal lethality character	Possible, e.g., any mechanical parts of the system, that directly or indirectly can cause the lethal outcome for human operator	N/A as lethal character depends on the expert level of the attacker
Qualitative criteria			
Expertise of threat actors	N/A	Essential, unless it is fully autonomous	Essential, but not suitable for developing agreements on level of expertise
Autonomy level	Automated and non-automated	Automated and autonomous	Automated and autonomous
Effects	Predictable	Predictable and unpredictable	Unpredictable
Capabilities	Constant parameters, development takes time, states have time to adjust	Constant parameters, development takes time, states have time to adjust	Evolving constantly; Agreements can be out-dated quickly
Implications	Clear	Clear and unclear in case of cyber attacks	Not always clear; Propagation can go out of control
Monitoring of compliance	Possible	Possible	Challenging, as requires intrusive access to governments networks; The access will provide the audit team with strategic information
External checks	Possible and passive	Possible and passive	Require active penetration

(Contd.)

Table 2.3: (*Contd.*)

Enforcement	Possible	Possible	Challenging because of attribution (anonymity, political reasons, reputation); Response and punishment (immediate or delayed, through cyber or conventional weapons, etc.)

One of the best examples might be a Chemical Sector Cyber Security Framework Implementation Guidance[198] published in 2015 by the Industrial Control Systems Cyber Emergency Response Team (ICS-CERT) for Improving Critical Infrastructure Cybersecurity developed by the National Institute of Standards and Technology (NIST).[199] Those guidelines aim to transfer cyber security measures for corporate networks to the chemical infrastructure and critical industrial networks. The NIST framework also provided comprehensive risk assessment and risk management solutions, recognized by the US Department of Homeland Security.

Cyber peace initiatives[200]

FifF, the German civil rights panel, has launched the Cyberpeace campaign to address the threats emerging from cyber warfare policies and to push back the colonization of the communication infrastructure by the military and surveillance of the entire population, which, in addition, sets everyone under suspicion.

Their goals are non-violent conflict resolution, arms control of cyber weapons and surveillance technology, dismissal of development and use of cyber weapons, the obligation to make IT vulnerabilities public and the promotion of communication infrastructure, which is, by law, secure against surveillance. FifF wants the Internet and for all infrastructure to be used in a peaceful fashion and to be protected against military misuse. FifF wants that secure communication can be ensured while preserving and promoting human and civil rights.

Cyber-physical arms, especially those of mass danger, can be assessed, monitored and controlled through bilateral and multilateral

---

[198] Chemical Sector Cybersecurity Framework Implementation Guidance, Homeland Security, 2015, last retrieved on 12 June 2018 at https://www.us-cert.gov/sites/default/files/c3vp/framework_guidance/chemical-framework-implementation-guide-2015-508.pdf

[199] Cybersecurity Framework, National Institute of Standards and Technology, last retrieved on 10 June 2018 at https://www.nist.gov/cyberframework

[200] S. Hugel, Cyberpeace Promoting human rights and peaceful use of the Internet, 2015, last retrieved on 15 August 2018 at https://sciforum.net/manuscripts/2978/manuscript.pdf

agreements, and further by a universally recognized international body. Preventive arms control is not necessarily limited to the prohibition of technical designs and components. In a wider sense, it can aim towards mutual understanding on the legitimate or illegitimate uses of emerging technological capacities—be it as a compliment or substitute to limitations on deployments, designs and numbers.[201] It is sufficient to ensure that the technological progress aimed at facilitating human lives is not jeopardized by any kind of ban out of the fear of dual use.

2.8 Legal and Ethical Cyber Issues

Cyberspace is an evolving virtual reality penetrating all aspects of our life, and it requires new approaches to regulating its internal and external environment and the establishment of a sound legal foundation. The challenging task is not only to understand the dual-use issues of legal, policy, ethical, strategic areas, but also to how regulate this emerging threat.

To confront these complex challenges, which have both national and international dimensions, and to develop concerted strategies, cyber security seeks advice and objective approaches from ethics and philosophy. In an ideal world, ethics is the primary foundation of law and public policy. Where law and policy are unclear it can be useful to return to "first principles" or basic ethical values to help clarify law and policy and to help illuminate the best path forward.[202]

Some basic ethical issues on the use of IT in global networks consists of ensuring personal privacy, data access rights, harmful actions in cyberspace, and confidence. These have been solved partially by using technological approaches, such as encryption techniques, digital IDs, cryptocurrency, firewalls, and intrusion detection systems. The average user would benefit from guidance on ethics, and experts would profit from legal regulations establishing the definition and limits of defence, for their rights and the responsibility of offence. International guidelines and strategies would be extremely helpful in ethical and socially sensitive access and use of globally accessible information.

Controversial issues
- What information about individuals can be revealed to others?
- Who is allowed to access critical data and information?

---

[201] Ibid
[202] UNIDIR Report on "The Weaponization of Increasingly Autonomous Technologies: Considering Ethics and Social Values", No. 3, UNIDIR, 2015, last retrieved on 19 October 2018 at http://www.unidir.ch/files/publications/pdfs/considering-ethics-and-social-values-en-624.pdf

- Does the launching of a cyberattack from one nation to another violate the neutrality of all the nations it traverses?
- Is it ethical to launch/use an offensive cyber weapon?
- If a criminal steals a computer from a legitimate cyber security expert, is the expert allowed to hack into his own computer to track the thief?
- If a cyber security expert is being blackmailed or "bullied" online, is he allowed to employ his skills to reveal the offender. What is the limit of cyber self-defence?
- Is use of a virus or worm unethical simply due to their uncontrollable nature?

Among the legal challenges cyber security faces today, the following are the most evident:

1. Lack of standard terminology (cyber arms, tools, weapons, intrusion, defence, etc.).
2. Lack of publicly available information and transparency for numerous cybe attacks, including on critical infrastructure.
3. Lack of jointly agreed international strategy and international regulations for cyberspace and cyber arms, including for conflicts.
4. Rapid development and acquisition of new generations of operational systems, cyber arms and weapons, including AWS implementations able to act without a human operator.

Legal and ethical cyber issues, going beyond countries, require a multi-sectoral and cross-country collaboration. Increasing connectivity did not only bring the prospect of human betterment; it created technological disparities between States and growing vulnerabilities and risks, or what are often described as "cyber insecurities" or "digital uncertainties". Such insecurities or uncertainties stem from the way different actors exploit ICT vulnerabilities, and the capacity of the targets to minimize the consequences and ensure business continuity. Consequences can be localized or, depending on the severity and the actors involved, escalate to the international level.[203]

Voluntary, Non-Binding Norms of Responsible State Behaviour in Peacetime
In the remainder of my remarks, I'd like to discuss very briefly another element of the United States' strategic framework for international cyber stability: the development of international consensus on certain additional voluntary, non-binding norms of responsible State behaviour in cyberspace that apply during peacetime. Internationally,

---

[203] C. Kavanagh, The United Nations, Cyberspace and International Peace and Security Responding to Complexity in the 21st Century, UNIDIR, 2017, last retrieved on 9 December 2018 at http://www.unidir.org/files/publications/pdfs/the-united-nations-cyberspace-and-international-peace-and-security-en-691.pdf (Reference made for "cyber insecurities": see M. Finnemore and D. Hollis, 2016. "Constructing Cyber Norms for Global Cyber Security", The American Journal of International Law, 110(3), pp. 425–447; and for "cyber uncertainties": see OECD (2015), "Digital Security Risk Management for Economic and Social Prosperity: OECD Recommendation and Companion Document")

the United States has identified and promoted four such norms:

- First, a State should not conduct or knowingly support cyber-enabled theft of intellectual property, trade secrets, or other confidential business information with the intent of providing competitive advantages to its companies or commercial sectors.
- Second, a State should not conduct or knowingly support online activity that intentionally damages critical infrastructure or otherwise impairs the use of critical infrastructure to provide service to the public.
- Third, a State should not conduct or knowingly support activity intended to prevent national computer security incident response teams (CSIRTs) from responding to cyber incidents. A State also should not use CSIRTs to enable online activity that is intended to do harm.
- Fourth, a State should cooperate, in a manner consistent with its domestic and international obligations, with requests for assistance from other States in investigating cybercrimes, collecting electronic evidence, and mitigating malicious cyber activity emanating from its territory.

These four U.S.-promoted norms seek to address specific areas of risk that are of national and/or economic security concern to all States. Although voluntary and nonbinding in nature, these norms can serve to define an international standard of behaviour to be observed by responsible, like-minded States with the goal of preventing bad actors from engaging in malicious cyber activity.[204]

(Brian J. Egan, International Law and Stability in Cyberspace, November 2016)[205]

The 2015 Companion Document to the OECD Recommendation of the Council on Digital Security Risk Management for Economic and Social Prosperity states that cybersecurity can be considered from four different (and often interdependent) perspectives:[206]

1. technology, which involves a focus on the actual digital environment (often called "information security", "computer security" or "network security" by experts);
2. law enforcement, and more generally, the legal framework (e.g. for responding to cybercrime);
3. national and international security (issues ranging from intelligence to the use of ICT in conflict); and
4. economic and social prosperity (relating to wealth creation, innovation, growth, competitiveness, and employment across economic sectors

---

[204] Brian J. Egan, International Law and Stability in Cyberspace, 35 Berkeley J. Int'l Law. 169, 2017, last retrieved on 7 December 2018 at http://scholarship.law.berkeley.edu/bjil/vol35/iss1/5

[205] Ibid

[206] OECD, Companion Document to the OECD Recommendation of the Council on Digital Security Risk Management for Economic and Social Prosperity, p. 19, 2015, last retrieved on 4 September 2018 at https://read.oecd-ilibrary.org/science-and-technology/digital-security-risk-management-for-economic-and-social-prosperity_9789264245471-en#page1

and other aspects such as individual liberties, health, education, culture, and so forth).

Strong efforts are undertaken by international organizations, unions and coalitions to help countries in establishing legal and normative grounds, assess their national capacities and address the gaps, harmonize legal frameworks, and be prepared to face the national and international implications of growing cyberthreats, and find mutually beneficial solutions in cyber conflicts.

Centre for the Governance of AI, Future of Humanity Institute, University of Oxford[207]
The Centre for the Governance of AI, housed at the Future of Humanity Institute, University of Oxford, strives to help humanity capture the benefits and mitigate the risks of artificial intelligence. Our focus is on the political challenges arising from transformative AI: advanced AI systems whose long-term impacts may be as profound as the industrial revolution. The Centre seeks to guide the development of AI for the common good by conducting research on important and neglected issues of AI governance, and advising decision makers on this research through policy engagement.

The Centre produces research which is foundational to the field of AI governance, for example mapping crucial considerations to direct the research agenda, or identifying distinctive features of the transition to transformative AI and corresponding policy considerations. Our research also addresses more immediate policy issues, such as malicious use and China's AI strategy. Current focuses include international security, the history of technology development, and public opinion.

2.8.1 International Regulations for Cyber Conflicts

On the international level conflicts of war are governed by the United Nations Charters, the Geneva Conventions, and the Hague Conventions. As these were developed in the pre-cyber era there is no common agreement or understanding on how they could be applied to the cyber domain. The technological evolution of cyber offence activities, especially in the AI realm, is no longer within the conventional conditions and existing international legal, ethical, and political frameworks.

The UN resolution A/RES/64/211 (2009) urged nations to acknowledge that trust and security when using information and communication technologies is a fundamental basis of the information society, and that it is necessary to stimulate, form, develop, and actively integrate a stable global culture of cybersecurity, and appealed for "Creation of a global culture of cybersecurity and taking stock of national efforts to protect critical information infrastructures".[208]

---

[207] Centre for the Governance of AI, Future of Humanity Institute, University of Oxford, last retrieved on 10 November 2018 at https://www.fhi.ox.ac.uk/governance-ai-program/

[208] Convention on International Information Security, The Ministry of Foreign Affairs of the Russian Federation, 22 September 2011, last retrieved on 14 December 2018 at http://www.mid.ru/en/foreign_policy/official_documents/-/asset_publisher/CptICk B6BZ29/content/id/191666

The international laws, set out in the UN Charter, while addressing the use of armed forces by states, considers two important questions related to any war—why the states fight and how they fight. The body of law known as *jus ad bellum* defines the legitimate reasons for a nation to use force against another nation and identifies certain criteria for a *just* war; while *jus in bello* regulates behaviour during armed conflicts and applies not only to governments and armed forces, but also to armed groups fighting each other.

Jus ad bellum, governed by Article 2(4) of the UN Charter, states that "All members shall refrain in their international relations from the threat or the use of force against the territorial integrity or political independence of any state, or in any other manner inconsistent with the purposes of the United Nations"; and Article 51: "Nothing in the present Charter shall impair the inherent right of individual or collective self-defence if an armed attack occurs against a Member of the United Nations", affirming the inherent right of states to self-defence in the case of an armed attack.

The rules governing the conduct of hostilities are set out in the Hague Conventions of 1899 and 1907. They limit the methods and means of warfare that parties to a conflict may use. In essence, they regulate the conduct of military operations in an armed conflict by defining proper and permissible uses of weapons and military tactics.[209] Theoretically, it is possible to break the rules while fighting a just war, or to be engaged in an unjust war while respecting the laws of armed conflict.

Opinions on the applicability of international law to the cyber reality vary. Thus, in a 2012 speech, Harold Koh, the then legal adviser to the US Secretary of State, explicitly stated the US view that the international law principles do apply in cyberspace.[210] In relation to the use of cyber weapons and *jus ad bellum* he noted that:

- cyber activities may in certain circumstances constitute uses of force within the meaning of Article 2(4) of the UN Charter and customary international law;
- a state's national right of self-defence, recognized in Article 51 of the UN Charter, may be triggered by computer network activities that amount to an armed attack or imminent threat thereof;
- states conducting activities in cyberspace must consider the sovereignty of other states, including outside the context of armed conflict; and

---

[209] Handbook on International Rules governing military operation, ICRC, 2013, last retrieved on 4 July 2018 at https://www.icrc.org/sites/default/files/topic/file_plus_list/0431-handbook_on_international_rules_governing_military_operations.pdf

[210] Harold Hongju Koh, International Law in Cyberspace, Remarks as Prepared for Delivery by Harold Hongju Koh to the USCYBERCOM Inter-Agency Legal Conference Ft. Meade, MD, Sept. 18, 2012, Harvard International Law Journal, Volume 54, December 2012, last retrieved on 14 July 2018 at https://digitalcommons.law.yale.edu/cgi/viewcontent.cgi?article=5858&context=fss_papers

- states are legally responsible for activities undertaken through "proxy actors" who act on the state's instructions or under its direction or control.

Regarding *jus in bello*, he said:

- in the context of an armed conflict, the law of armed conflict applies to regulate the use of cyber tools in hostilities, just as it does other tools;
- the *jus in bello* principle of distinction (that is, distinguishing between military and non-military objectives) applies to computer network attacks undertaken in the context of an armed conflict; and
- the *jus in bello* principle of proportionality applies to computer network attacks undertaken in the context of an armed conflict.

In 2016, Brian J. Egan, the US State Department Legal Adviser, offered some additional US views on how certain rules of international law apply to States' behaviours in cyberspace. He noted that if cyber operations constitute "attacks" under the law of armed conflict, the rules on conducting attacks must be applied to those cyber operations. Such operations must only be directed against military objectives and comport with the requirements of the principles of distinction and proportionality. For increased transparency, States need to do more work to clarify how the international law on non-intervention applies to States' activities in cyberspace. In an ideal scenario, the Internet must remain open to the free flow of information and ideas and the concept of State sovereignty should not be used as a justification for excessive regulation of online content, including censorship and access restrictions, often undertaken in the name of counterterrorism or countering violent extremism. Any regulation by a State of matters within its territory, including the use of and access to the Internet, must comply with that State's applicable obligations under international human rights law. [211]

Although many States, including the United States, generally believe that the existing international legal framework is sufficient to regulate State's behaviour in cyberspace, States likely have divergent views on specific issues. Further discussion, clarification, and cooperation on these issues remains necessary.[212]

The following reports and proposals should be mentioned to highlight achievements in the development of the cyber legal environment at the international level.

The Tallinn Manuals

The Tallinn Manual on International Law Applicable to Cyber Warfare

---

[211] Brian J. Egan, International Law and Stability in Cyberspace, 35 Berkeley J. Int'l Law. 169, April 2017, last retrieved on 14 July 2018 at http://scholarship.law.berkeley.edu/bjil/vol35/iss1/5

[212] Ibid

of 2013 presented the views of twenty international law scholars and practitioners on how international law applies to cyber warfare and proposes ninety-five "black-letter rules" relevant to cyber conflict that can be derived from international law (including law related to sovereignty, state responsibility, and neutrality, as well as the UN Charter and the Geneva Conventions).[213]

The Tallinn Manual on the International Law Applicable to Cyber Warfare
Rule 9 states, "A state injured by an internationally wrongful act may resort to proportionate countermeasures, including cyber countermeasures, against the responsible state".
Rule 10 states, "A cyber operation that constitutes a threat or use of force against the territorial integrity or political independence of any State, or that is in any other manner inconsistent with the purposes of the United Nations, is unlawful".[214]
Rule 37 states, "Civilian objects shall not be made the object of cyberattacks. Computers, computer networks, and cyber infrastructure may be made the object of attack if they are military objectives".[215]

The second edition, Tallinn Manual 2.0 on the International Law Applicable to Cyber Operations, updated and considerably expanded the 2013 Tallinn Manual on the International Law Applicable to Cyber Warfare. It added legal analysis of most common cyber incidents and confirmed that the pre-cyber international law applies to cyber operations, and that the states have both rights and obligations under international law. [216]

As such, the 2017 edition covers a full spectrum of international law as applicable to cyber operations, ranging from peacetime legal regimes to the law of armed conflict. The analysis of a wide array of international law principles and regimes that regulate events in cyber space include principles of general international law, such as the sovereignty and the various bases for the exercise of jurisdiction. The law of state responsibility, which includes the legal standards for attribution, is examined at length. Additionally, numerous specialized regimes of international law, including human rights law, air and space law, the law of the sea, and diplomatic and consular law are examined within the context of cyber operations."[217]

The United Nations Activities

Within the United Nations, the Member States are developing a normative base to shape the behaviour of different actors (States, criminal actors,

---

[213] Michael N. Schmitt et al., eds., 2013, Tallinn Manual on the International Law Applicable to Cyber Warfare, Cambridge University Press, Cambridge UK.

[214] Ibid., 42

[215] Ibid., 124

[216] Tallinn Manual 2.0 on the International Law Applicable to Cyber Operations, last retrieved on 14 October 2018 at https://ccdcoe.org/sites/default/files/documents/CCDCOE_Tallinn_Manual_Onepager_web.pdf

[217] Ibid

users, etc.) in both peacetime and in the context of armed conflict in their use of cyberspace and ICTs, thereby ensuring a stable ICT environment. From a political–military perspective, this policy option has included the work of the General Assembly First Committee on Disarmament and International Security, which, through its successive Groups of Governmental Experts (GGEs) on Developments in the Field of Information and Telecommunications in the Context of International Security, has facilitated some of the first efforts to reach global consensus on the binding and non-binding norms that apply to the digital environment and the behaviour of States in their uses of ICT. [218]

The UN Group of Governmental Experts (GGE) on Information, Telecommunications, and International Security

The GGE activity was launched in 2004. This group, comprised of governmental IT experts from fifteen nations (Argentina, Australia, Belarus, Canada, China, Egypt, Estonia, France, Germany, India, Indonesia, Japan, the Russian Federation, the United Kingdom, and the United States), and was established at the request of the UN General Assembly to study "existing and potential threats in the sphere of information security and possible cooperative measures to address them including norms, rules or principles of responsible behaviour of States and confidence-building measures with regard to the information space, as well as the concepts aimed at strengthening the security of global information and telecommunications systems".[219] By 2016 that number of countries in the group had grown to 25.

The UN GGE can be credited with two major achievements: outlining the global cybersecurity agenda and introducing the principle that international law applies to the digital space.[220]

The 2013 GGE report concludes that "International law, and in particular the Charter of the United Nations, is applicable and is essential to maintaining peace and stability and promoting an open, secure,

---

[218] Camino Kavanagh, The United Nations, Cyberspace and International Peace and Security Responding to Complexity in the 21st Century, UNIDIR, 2017, last retrieved on 9 December 2018 at http://www.unidir.org/files/publications/pdfs/the-united-nations-cyberspace-and-international-peace-and-security-en-691.pdf

[219] "Statement by the Chair of the United Nations Group of Governmental Experts on Developments in the Field of Information and Telecommunications in the Context of International Security, H.E. Ambassador Deborah Stokes of Australia," 25 October 2013, last retrieved on 14 July 2018 at http://www.un.org/disarmament/special/meetings/firstcommittee/68/pdfs/TD_25-Oct_OWMD_Chair_UNGGE.pdf

[220] UN GGE, Digital Watch, last retrieved on 14 July 2018 at https://dig.watch/processes/ungge

peaceful and accessible ICT environment".[221] Although this statement is a recommendation of the group of experts rather than an authoritative commitment from the nations they represent, many of the experts in the group have formal affiliations with their national governments and this has frequently been interpreted as indicating a concurrence among the represented nations that international law applies to cyberspace. The report was presented to the UN General Assembly, … after which the General Assembly unanimously took note of the report without accepting any of its specific assessments or recommendations.[222]

In addition, the 2013 report, A/68/98[223], recognized the dual-use nature of ICT as either legitimate or malicious, the security challenge posed by global interconnectedness and anonymity, and the hostile potential of state and non-state actors. It also emphasized responsible state behaviour, particularly state sovereignty and jurisdiction over ICT infrastructure, as well as states' responsibilities for proxies and operation of non-State actors within national jurisdiction.[224]

In July 2015, a new UN group of government experts issued a second report on "Developments in the Field of Information and Telecommunications in the Context of International Security".[225] The new group numbered 20 countries and included Antigua and Barbuda, Belarus, Brazil, China, Colombia, Egypt, Estonia, France, Germany, Ghana, Israel, Japan, Kenya, Malaysia, Mexico, Pakistan, Russia, Spain, the United Kingdom, and the United States.

Although the 2015 report does explicitly endorse other parts of the 2013 report, such as those related to capacity building, it does not explicitly endorse the conclusion of the 2013 report that the international law is applicable to cyber space. However, Paragraph 28(c) of the 2015 report says, "Underscoring the aspirations of the international community to

---

[221] Group of Governmental Experts on Developments in the Field of Information and Telecommunications in the Context of International Security, A/68/98, United Nations General Assembly, 24 June 2013, last retrieved on 14 August 2018 at http://www.un.org/ga/search/view_doc.asp?symbol=A/68/98

[222] Developments in the Field of Information and Telecommunications in the Context of International Security, A/RES/68/243, United Nations General Assembly, 9 January 2014, last retrieved on 14 August 2018 at http://www.un.org/ga/search/view_doc.asp?symbol=A/RES/68/243.

[223] Group of Governmental Experts on Developments in the Field of Information and Telecommunications in the Context of International Security, A/68/98, United Nations, 68th sess, 2013, last retrieved on 19 August 2018 at http://www.unidir.org/files/medias/pdfs/developments-in-the-field-of-information-and-telecommunications-in-the-context-of-international-security-2012-2013-a-68-98-eng-0-518.pdf

[224] Ibid

[225] Group of Governmental Experts on Developments in the Field of Information and Telecommunications in the Context of International Security, A/70/174, United Nations General Assembly, 22 July 2015, last retrieved on 16 November 2018 at http://www.un.org/ga/search/view_doc.asp?symbol=A/70/174

the peaceful use of ICTs for the common good of mankind, and recalling that the Charter applies in its entirety, the Group noted the inherent right of States to take measures consistent with international law and as recognized in the Charter."[226]

Three of those GGEs (2010[227], 2013[228], and 2015[229]) led to consensus reports that recommended states abide by a set of norms—including the applicability of international law to cyberspace, participate in confidence building measures, and support capacity building initiatives to reduce the risk that state actions in cyberspace threaten international peace and security.[230] However, the norms and rules proposed in the reports remained mostly the conclusions and recommendation of experts since they were only advisory.

In 2016, the General Assembly adopted resolution 71/28[231] on developments in the field of information and telecommunications in the context of international security. It welcomed the GGE conclusions of the 2013 and 2015 reports, that international law, and in particular the Charter of the United Nations, is applicable and essential to maintaining peace and stability and promoting an open, secure, stable, accessible and peaceful information and communications technology environment.

In 2016-2017, several draft substantive reports were considered by the Group on the following issues: existing and emerging threats; capacity-building; confidence-building; recommendations on the implementation of norms, rules and principles for the responsible behaviour of States; application of international law to the use of information and communications technologies; and conclusions and recommendations

---

[226] Ibid

[227] Group of Governmental Experts on Developments in the Field of Information and Telecommunications in the Context of International Security, A/65/201, United Nations General Assembly, 30 July 2010, last retrieved on 19 November 2018 at https://undocs.org/A/65/201

[228] Group of Governmental Experts on Developments in the Field of Information and Telecommunications in the Context of International Security, A/68/98*, United Nations General Assembly, 24 June 2013, last retrieved on 19 November 2018 at http://www.un.org/ga/search/view_doc.asp?symbol=A/68/98

[229] Group of Governmental Experts on Developments in the Field of Information and Telecommunications in the Context of International Security, A/70/174, United Nations General Assembly, 22 July 2015, last retrieved on 19 November 2018 at http://www.un.org/ga/search/view_doc.asp?symbol=A/70/174

[230] A. Grisby, Unpacking the Competing Russian and U.S. Cyberspace Resolutions at the United Nations, Council on Foreign Relations, 29 October 2018, last retrieved on 19 November 2018 at https://www.cfr.org/blog/unpacking-competing-russian-and-us-cyberspace-resolutions-united-nations

[231] Resolution: Developments in the field of information and telecommunications in the context of international security, United Nations General Assembly, A/RES/71/28, 9 December 2016, last retrieved on 19 November 2018 at https://undocs.org/A/RES/71/28

for future work. The consensus was not reached.[232] Self-defence, countermeasures and attribution seem to be the most controversial areas in achieving the overall agreement.

The 2017 UNIDIR Report "The United Nations, Cyberspace and International Peace and Security Responding to Complexity in the 21st Century" states that despite many positive developments and the gradual spread of norms, the foundations for establishing a strong normative framework around the use of ICTs in the context of international peace and security appear to be faltering.[233]

The development of strong normative international cyber framework is complicated by several overlapping factors:
- persisting disagreements among States on how existing rules of international law apply to State uses of ICTs. The sources of these disagreements—many of which spill over into other policy areas—lie just as much in different appreciations of the legal issues at hand as they do in perceptions of strategic imbalances in military and civilian IT resources and capabilities and were—to a large degree—what prevented the 2016–2017 GGE from agreeing on a consensus report;
- the sluggishness of some States in moving beyond mere process to practical implementation of the recommended norms of State behaviour;
- an equally serious lack of capacity and resources to implement some of the recommended norms and confidence-building measures, including for establishing the national structures and mechanisms required to respond to ICT vulnerabilities and risks, and for attributing ICT-related incidents;
- a lack of awareness by policymakers in many States of the different normative processes underway within and beyond the UN relating directly or indirectly to international peace and security; and
- a deepening lack of trust among various stakeholders, driven in part by broader, non-technological issues affecting relations among States, and which undermines collaboration and cooperation. If States are sincere in their commitments, they need to strengthen the normative foundations for responsible behaviour in the use of ICTs across policy agendas. This can be achieved by shifting to practical implementation of those norms and measures that have already been agreed upon, helping bridge existing capacity needs, and by ensuring that existing multilateral and bilateral channels are kept open for continued dialogue on those issues on which agreement is less tangible in the immediate term.[234]

Source: The 2017 UNIDIR Report
The United Nations, Cyberspace and International Peace
and Security Responding to Complexity in the 21st Century

---

[232] Fact Sheet: Developments in the Field of Information and Telecommunications in the Context of International Security, United Nations Office for Disarmament Affairs, last retrieved on 24 November 2018 https://s3.amazonaws.com/unoda-web/wp-content/uploads/2018/07/Information-Security-Fact-Sheet-July2018.pdf

[233] C. Kavanagh, The United Nations, Cyberspace and International Peace and Security Responding to Complexity in the 21st Century, UNIDIR, 2017, last retrieved on 9 December 2018 at http://www.unidir.org/files/publications/pdfs/the-united-nations-cyberspace-and-international-peace-and-security-en-691.pdf

[234] Ibid

In the absence of overall international agreements, states move more towards bilateral and multilateral agreements, a trend which was particularly prevalent in 2015 and 2016.

In 2015, the US and China signed the Memorandum of Understanding on US-China Development Cooperation and the Establishment of an Exchange and Communication Mechanism between the United States Agency for International Development and the Ministry of Commerce of the People's Republic of China. They agreed that timely responses should be provided to requests for information and assistance concerning malicious cyber activities. Further, both sides agreed to cooperate, in a manner consistent with their respective national laws and relevant international obligations, with requests to investigate cybercrimes, collect electronic evidence, and mitigate malicious cyber activity emanating from their territory. Both sides also agreed that neither country's government will conduct or knowingly support cyber-enabled theft of intellectual property, including trade secrets or other confidential business information, with the intent of providing competitive advantages to companies or commercial sectors. Both sides are committed to making common efforts to further identify and promote appropriate norms of state behaviour in cyberspace within the international community.[235]

The same year, the leaders of the G20, the world's largest advanced and emerging economies, agreed in the communiqué from their summit in Antalya, Turkey, that no country should conduct or support cyber theft of intellectual property, and that "we [the leaders of the G20] affirm that international law, and in particular the UN Charter, is applicable to state conduct in the use of ICTs".[236]

Among other notable multi-country activities, there is the Budapest Convention and the Shanghai Cooperation Organization.

The Budapest Convention

The Budapest Convention on Cybercrime in 2001 is an international agreement among forty-seven nations (including most members of the Council of Europe, the United States, Canada, Australia, and Japan).[237]

---

[235] White House, Office of the Press Secretary, "FACT SHEET: President Xi Jinping's State Visit to the United States," 25 September 2015, last retrieved on 13 September 2018 at https://www.whitehouse.gov/the-press-office/2015/09/25/fact-sheet-president-xi-jinpings-state-visit-united-states

[236] The G20 members are Argentina, Australia, Brazil, Canada, China, France, Germany, India, Indonesia, Italy, Japan, Republic of Korea, Mexico, Russia, Saudi Arabia, South Africa, Turkey, the United Kingdom, the United States, and the European Union, last retrieved on 13 September 2018 at http://g20.org.tr/about-g20/g20-members/

[237] Chart of Signatures and Ratifications of Treaty 185: Convention on Cybercrime, Council of Europe, updated 11 February 2016, last retrieved on 13 September 2018 at http://conventions.coe.int/Treaty/Commun/ChercheSig.asp?NT=185&CM=&DF=&CL=ENG

The convention has three main purposes: to enact domestic laws that criminalize certain kinds of behaviour in cyberspace; to implement certain investigative procedures for law enforcement in the signatory nations; and to enhance international cooperation regarding law enforcement activities against cybercrime. The Budapest Convention itself does not establish international law that criminalizes specific behaviours. Rather, it aims to harmonise the domestic criminal law across the signatory nations regarding these behaviours. Russia and China are not parties to the convention.

The Agreement between the Member States of the Shanghai Cooperation Organization

In June 2009, the six member states of the Shanghai Cooperation Organization (Russia, China, Kazakhstan, Kyrgyzstan, Tajikistan, and Uzbekistan) concluded a Russian-drafted agreement defining "information wars" broadly as a "confrontation between two or more states in the information space aimed at damaging information systems, processes and resources, critical and other structures, undermining political, economic and social systems, [and] mass psychologic brainwashing to destabilize society and state".[238]

For the most part, this agreement was a joint statement among the signatories emphasizing their views on the undesirability of information war. However, Article 4(1) states, "The parties shall cooperate and act in the international information space within the framework of this agreement in such a way that the activities contribute to social and economic development and comply with maintaining international security and stability, generally recognized principles and norms of international law, including the principles of peaceful settlement of disputes and conflicts, non-use of force, non-interference in internal affairs, respect for human rights and fundamental freedoms and the principles of regional cooperation and non-interference in the information resources of the States of the Parties".

This statement does not require any signatory to refrain from any particular action in cyberspace. The agreement commits the parties to take actions that contribute to social and economic development in a manner consistent with the principles listed, but it does not explicitly prohibit one signatory from launching cyberattacks against another if, in the judgment

---

[238] "Agreement between the Governments of the Member States of the Shanghai Cooperation Organization on Cooperation in the Field of International Information Security, Yekaterinburg, 16 June 2009," in S. A. Komov, ed., International Information Security: The Diplomacy of Peace—Compilation of Publications and Documents (Moscow, 2009), 202–213. This is an unofficial translation; the authentic languages of the Agreement are Russian and Chinese.

of the launching nation, such attacks might help to maintain security and stability.[239]

In 2015, the Member States of the Shanghai Cooperation Organization submitted to the General Assembly of the United Nations the International Code of Conduct for Information Security. The paragraphs 2(7) (protecting online and offline rights equally) and 2(8) (facilitating access for all in cyberspace), presented new opportunities for countries to bridge certain gaps in setting cyber norms.

In October 2018, two competing draft resolutions were submitted to the UN General Assembly for consideration. As expected, Russia tabled a draft resolution seeking the General Assembly's endorsement of an "international code of conduct for international information security," and a resumption of the UN Group of Governmental Experts (GGE) process next year.... The United States tabled a competing resolution, setting up a clash between Russia, China, and their largely autocratic friends on one side, and the United States, the European Union, Canada, Japan, and Australia on the other.[240]

Countries supporting two draft resolutions submitted to the UN General Assembly in 2018
Draft resolution A/C.1/73/L.27[241]
Developments in the field of information and telecommunications in the context of international security
Algeria, Angola, Azerbaijan, Belarus, Bolivia (Plurinational State of), Burundi, Cambodia, China, Cuba, Democratic People's Republic of Korea, Democratic Republic of the Congo, Eritrea, Kazakhstan, Madagascar, Namibia, Nepal, Nicaragua, Pakistan, Russian Federation, Samoa, Sierra Leone, Suriname, Syrian Arab Republic, Tajikistan, Uzbekistan, Venezuela (Bolivarian Republic of) and Zimbabwe
Draft Resolution A/C.1/73/L.37[242]
Advancing responsible State behaviour in cyberspace in the context of international security
Australia, Austria, Belgium, Bulgaria, Canada, Croatia, Cyprus, Czechia, Denmark, Estonia, Finland, France, Georgia, Germany, Greece, Hungary, Ireland, Israel, Italy, Japan, Latvia, Lithuania, Luxembourg, Malawi, Malta, Netherlands, Poland, Portugal, Romania, Slovakia, Slovenia, Spain, Sweden, Ukraine, United Kingdom of Great Britain and Northern Ireland and United States of America

---

[239] E.D .Harris, J.M. Acton, H .Lin, Governance of Dual-Use Technologies: Theory and Practice, April 2016, last retrieved on 13 September 2018 at https://www.amacad.org/content/publications/pubContent.aspx?d=22234#toNote31

[240] Unpacking The Competing Russian and U.S. Cyberspace Resolutions at the United Nations, 29 October 2018, last retrieved on 2 December 2018 at https://www.cfr.org/blog/unpacking-competing-russian-and-us-cyberspace-resolutions-united-nations

[241] Draft Resolution A/C.1/73/L.27, Developments in the field of information and telecommunications in the context of international security, at https://undocs.org/A/C.1/73/L.27

[242] Draft Resolution A/C.1/73/L.37, Advancing responsible State behaviour in cyberspace in the context of international security, at http://undocs.org/A/C.1/73/L.37

All the above shows is that the international community through dialogues, discussions, even with contradictions, is moving forward in exploring various options for the development of universal cyber laws. It will take time and more joint efforts to succeed.

2.8.2 Against Autonomous Weapon Systems for Military

The AI-enhanced cyber arms and autonomous weapons systems pose new threats and challenges in legal regulations by autonomously selecting their targets. Production and acquisition of AWS for military use is widely discussed by the concerned international community, leading cyber experts, governing bodies and international organizations. The most evident and simplest solution to ban their development is not applicable in this case as it will negatively impact technological progress.

Autonomous systems designed to cause physical harm have additional ethical dimensions as compared to both traditional weapons and autonomous systems not designed to cause harm. Multi-year discussions on international legal agreements around autonomous systems in the context of armed conflict are led by the United Nations (UN), but professional ethics about such systems can and should have ethical standards covering a broad array of issues arising from the automated targeting and firing of weapons.[243]

The UNIDIR paper on "The Weaponization of Increasingly Autonomous Technologies: Considering Ethics and Social Values" (2015) highlights some of the ethical and social issues that arise from the discussion on AWSs. It suggests that far from being "extraneous" to the policy debate on the weaponization of increasingly autonomous technologies, ethics and social values are close to the core of this discussion. Although legal and technical discussions may produce information about possible technological trajectories, future applications and rules, they will not necessarily produce the insights, wisdom and prudence needed for an established advisory policy that will serve national and international interests. [244]

The UN Report of the Special Rapporteur on extrajudicial, summary or arbitrary executions, Christof Heyns to the UN Human Rights Council, 2013, states that lethal autonomous robots (LARs) "raise far-reaching concerns about the protection of life during war and peace. This includes the question of the extent to which they can be programmed to

---

[243] Reframing Autonomous Weapon Systems, The IEEE Global Initiative on Ethics of Autonomous and Intelligent Systems, *Ethically Aligned Design*, Version 2, IEEE, 2017, last retrieved on 8 October 2018 at https://standards.ieee.org/content/dam/ieee-standards/standards/web/documents/other/ead_reframing_autonomous_weapons_v2.pdf

[244] "The Weaponization of Increasingly Autonomous Technologies: Considering Ethics and Social Values", UNIDIR, 2015, last retrieved on 19 October 2018 at http://www.unidir.ch/files/publications/pdfs/considering-ethics-and-social-values-en-624.pdf

comply with the requirements of international humanitarian law and the standards protecting life under international human rights law. Beyond this, their deployment may be unacceptable because no adequate system of legal accountability can be devised, and because robots should not have the power of life and death over human beings." The Special Rapporteur recommends that States establish national moratoria on aspects of LARs and calls for the establishment of a high-level panel on LARs to articulate a policy for the international community on the issue.[245]

The below initiatives illustrate the strong concerns expressed by the global community about the danger of military application of AI and LAWSs.

In 2013, the International Committee for Robot Arms Control, a founder of the Campaign to Stop Killer Robots, issued a statement endorsed by more than 270 engineers, computing and artificial intelligence experts, roboticists, and professionals from related disciplines that called for a ban on fully autonomous weapons. In the statement, 272 experts from 37 countries say that, "given the limitations and unknown future risks of autonomous robot weapons technology, we call for a prohibition on their development and deployment. Decisions about the application of violent force must not be delegated to machines."[246]

Open Letter to ban warfare AI and Autonomous Weapons

In July 2015, over 1,000 high-profile artificial intelligence experts and leading researchers signed an open letter warning of a "military artificial intelligence arms race" and calling for a ban on "offensive autonomous weapons".

The letter, presented at the International Joint Conference on Artificial Intelligence in Buenos Aires, Argentina, was signed by Tesla's Elon Musk, Apple co-founder Steve Wozniak, Google DeepMind chief executive Demis Hassabis and Professor Stephen Hawking along with 1,000 AI and robotics researchers.

The letter states: "AI technology has reached a point where the deployment of [autonomous weapons] is—practically if not legally—feasible within years, not decades, and the stakes are high: autonomous weapons have been described as the third revolution in warfare, after gunpowder and nuclear arms." The authors argue that AI can be used to make the battlefield a safer place for military personnel, but that offensive weapons that operate on their own would lower the threshold of going to battle and result in a greater loss of human life.

Should one military power start developing systems capable of selecting targets and operating autonomously without direct human control, it would start an arms race similar to the one for the atom bomb, the authors argue. Unlike nuclear weapons,

---

[245] Report of the Special Rapporteur on extrajudicial, summary or arbitrary executions, Christof Heyns, United Nations General Assembly, A/HRC/23/47, 9 April 2013, last retrieved on 9 October 2018 at https://www.ohchr.org/Documents/HRBodies/HRCouncil/RegularSession/Session23/A-HRC-23-47_en.pdf

[246] Scientists call for a ban, Campaign to stop killer robots, last retrieved on 8 July 2018 at https://www.stopkillerrobots.org/2013/10/scientists-call/

> however, AI requires no specific hard-to-create materials and will be difficult to monitor.
>
> "The endpoint of this technological trajectory is obvious: autonomous weapons will become the Kalashnikovs of tomorrow. The key question for humanity today is whether to start a global AI arms race or to prevent it from starting," said the authors.[247]

In August 2017, an Open Letter signed by 160 CEOs was addressed to the United Nations Convention on Certain Conventional Weapons[248] urging the UN to double their efforts.

> **An Open Letter to the United Nations Convention on Certain Conventional Weapons**
>
> As companies building the technologies in Artificial Intelligence and Robotics that may be
>
> Re-purposed to develop autonomous weapons, we feel especially responsible in raising this alarm.
>
> We warmly welcome the decision of the UN's Conference of the Convention on Certain Conventional Weapons (CCW) to establish a Group of Governmental Experts (GGE) on Lethal Autonomous Weapon Systems. Many of our researchers and engineers are eager to offer technical advice to your deliberations...
>
> We entreat the High Contracting Parties participating in the GGE to work hard at finding means to prevent an arms race in these weapons, to protect civilians from their misuse, and to avoid the destabilizing effects of these technologies.
>
> We regret that the GGE's first meeting, which was due to start today, has been cancelled due to a small number of states failing to pay their financial contributions to the UN. We urge the High Contracting Parties therefore to double their efforts at the first meeting of the GGE now planned for November.
>
> Lethal autonomous weapons threaten to become the third revolution in warfare. Once developed, they will permit armed conflict to be fought at a scale greater than ever, and at timescales faster than humans can comprehend. These can be weapons of terror, weapons that despots and terrorists use against innocent populations, and weapons hacked to behave in undesirable ways.
>
> We do not have long to act. Once this Pandora's box is opened, it will be hard to close.
>
> We therefore implore the High Contracting Parties to find a way to protect us all from these dangers.

In 2016, the Fifth Review Conference of the High Contracting Parties to the Convention on Certain Conventional Weapons (CCW) established a Group of Governmental Experts (GGE) on emerging technologies in the

---

[247] Samuel Gibbs, Musk, Wozniak and Hawking urge ban on warfare AI and autonomous weapons, 27 July 2015, last retrieved on 8 October 2018 at https://www.theguardian.com/technology/2015/jul/27/musk-wozniak-hawking-ban-ai-autonomous-weapons

[248] An Open Letter to the United Nations Convention on Certain Conventional Weapons, 2017, last retrieved on 11 October 2018 at https://futureoflife.org/autonomous-weapons-open-letter-2017/?cn-reloaded=1

area of lethal autonomous weapons systems (LAWS). The GGE's mandate is to examine emerging technologies in the area of LAWS, in the context of the objectives and purposes of the CCW, and with a view toward identification of the rules and principles applicable to such weapon systems.

The overarching issues in the area of LAWS that are addressed in the 2018 meetings of the GGE include:[249]

1. Characterization of the systems under consideration in order to promote a common understanding on concepts and characteristics relevant to the objectives and purposes of the CCW;
2. Further consideration of the human element in the use of lethal force; aspects of human-machine interaction in their development, deployment and use of emerging technologies in the area of lethal autonomous weapons systems;
3. Review of potential military applications of related technologies in the context of the Group's work;
4. Possible options for addressing the humanitarian and international security challenges posed by emerging technologies in the area of LAWS in the context of the objectives and purposes of the Convention without prejudging policy outcomes and considering past, present and future proposals.

At the April 2018 GGE meeting, four more countries joined the call for the ban of LAWS, making it 26 countries in total.[250] France, Israel, Russia, United Kingdom, and United States explicitly rejected any move to negotiate new international laws on fully autonomous weapons. Approximately two dozen mostly European states indicated that they see no need to conclude new international law at this time.

Countries that have called for a ban on LAWS (April 2018)
(In alphabetical order)

Algeria	Ghana
Argentina	Guatemala
Austria	Holy See
Bolivia	Iraq

---

[249] 2018 Group of Governmental Experts on Lethal Autonomous Weapons Systems (LAWS), The United Nations Office at Geneva, last retrieved on 11 October 2018 at https://www.unog.ch/80256EE600585943/(httpPages) /7C335E71DFCB29D1C1258243003E8724? Open Document

[250] Report on Activities, 9-13 April 2018, Campaign to stop killer robots, last retrieved on 8 May 2018 at https://www.stopkillerrobots.org/wp-content/uploads/2018/07/KRC_ReportCCWX_Apr2018_UPLOADED.pdf

Brazil	Mexico
Chile	Nicaragua
China	Pakistan
Colombia	Panama
Costa Rica	Peru
Cuba	State of Palestine
Djibouti	Uganda
Ecuador	Venezuela
Egypt	Zimbabwe

The 2018 UNIDIR paper on "The Weaponization of Increasingly Autonomous Technologies: Artificial Intelligence"[251] continues the international discussions on the weaponization of increasingly autonomous technologies and states that "because of the complex legal, moral, ethical, and other issues raised by AI systems, policymakers are best served by a cross disciplinary dialogue that includes scientists, engineers, military professionals, lawyers, ethicists, academics, members of civil society, and other voices. Including diverse perspectives in on-going discussions surrounding lethal autonomous weapons that can help ensure that militaries use emerging technologies in responsible ways.

> The IEEE Global Initiative on Ethics of Autonomous and Intelligent Systems recommends that technical organizations promote a number of measures to help ensure that there is meaningful human control of weapons systems:
> - That automated weapons have audit trails to help guarantee accountability and control.
> - That adaptive and learning systems can explain their reasoning and decisions to human operators in transparent and understandable ways.
> - That there be responsible human operators of autonomous systems who are clearly identifiable.
> - That the behaviour of autonomous functions should be predictable to their operators.
> - That those creating these technologies understand the implications of their work.
> - That professional ethical codes are developed to appropriately address the development of autonomous systems and autonomous systems intended to cause harm.
>
> Specifically, we would like to ensure that stakeholders are working with sensible and comprehensive shared definitions, particularly for key concepts relevant to autonomous weapons systems (AWS). Designers should always ensure their designs

---

[251] "The Weaponization of Increasingly Autonomous Technologies: Artificial Intelligence" UNIDIR, 2018, last retrieved on 19 December 2018 at http://www.unidir.ch/files/publications/pdfs/the-weaponization-of-increasingly-autonomous-technologies-artificial-intelligence-en-700.pdf

meet the standards of international humanitarian law, international human rights law, and any treaties or domestic law of their particular countries, as well as any applicable engineering standards, military requirements, and governmental regulations. We recommend designers not only take stands to ensure meaningful human control but be proactive about providing quality situational awareness to operators and commanders using those systems. Professional ethical codes should be informed by not only the law, but an understanding of both local- and global-level ramifications of the products and solutions developed. This should include thinking through the intended use or likely abuse that can be expected by users of AWS.[252]

Those against the development of LAWS state that such weapons would violate the basic human rights and threaten the right to life and principle of preserving human dignity. "Once available in the military arsenal, LAWS would be perfect tools of repression in the hands of a few developed States."[253]

In July 2018, tech leaders, including Elon Musk and the three co-founders of Google's AI subsidiary DeepMind, signed a pledge promising to not develop "lethal autonomous weapons."[254]

Granting the life-or-death decision making to machines, allowing them to choose targets is one of the issues of the largest concern for those who are trying to ban LAWSs. The distinction between combatant and non-combatant might not be so easy to make, as well as to further identify those responsible for civilian deaths (a fundamental condition of *jus in bello*) and to determine of accountability. Potential cyber threats, even in the worst-case scenario of AI against AI battle, may lead to implementation of incorrect commands and false counterattacks, etc.

Report on Activities
Convention on Conventional Weapons Group of Governmental Experts meeting on lethal autonomous weapons systems
United Nations Geneva
9–13 April 2018

A total of 84 countries participated in the April 2018 GGE meeting. Other participants included UN agencies such as UNIDIR, the International Committee of the Red Cross (ICRC), Campaign to Stop Killer Robots, and various academics.

---

[252] Reframing Autonomous Weapon Systems, The IEEE Global Initiative on Ethics of Autonomous and Intelligent Systems, Ethically Aligned Design, Version 2, IEEE, 2017, last retrieved on 8 October 2018 at https://standards.ieee.org/content/dam/ieee-standards/standards/web/documents/other/ead_reframing_autonomous_weapons_v2.pdf

[253] Dr U C Jha (Retd) (2016) Killer Robots: Lethal Autonomous Weapon Systems Legal, Ethical and Moral Challenges

[254] J. Vincent, "Elon Musk, DeepMind founders, and others sign pledge to not develop lethal AI weapon systems", 18 July 2018, The Verge, last retrieved 20 July 2018 at https://www.theverge.com/2018/7/18/17582570ai-weapons-pledge-elon-musk-deepmind-founders-future-of-life-institute

This marked the fifth CCW meeting on lethal autonomous weapons systems since 2014 and the second of the Group of Governmental Experts, established in 2016.

During the meeting, there was significant convergence on the urgent need to negotiate new international law to retain meaningful human control over weapons systems and the use of force. For example, Austria and other states proposed that the CCW agree in November to a mandate to negotiate a legally-binding instrument (i.e. protocol or treaty) on fully autonomous weapons.

A group of African states recommended concluding a legally binding instrument "at the earliest" and found that "fully autonomous weapons systems or LAWS that are not under human control should be banned."

A working paper submitted to the meeting by the Non-Aligned Movement called for a "legally binding international instrument stipulating prohibitions and regulations on lethal Autonomous weapons systems."

The campaign invites national statements affirming support for these objectives expressed in group statements.

Virtually all states that spoke during the meeting stressed the necessity of retaining human control over weapons systems and the use of force.[255]

In September 2018, the European Parliament passed a resolution calling for an international ban on lethal autonomous weapons systems (LAWS).[256] Among other things, the resolution calls on its Member States and the European Council "to develop and adopt, as a matter of urgency … a common position on lethal autonomous weapon systems that ensures meaningful human control over the critical functions of weapon systems, including during deployment." The resolution also urges Member States and the European Council "to work towards the start of international negotiations on a legally binding instrument prohibiting lethal autonomous weapons systems."

- Governments are responsible for the regulation of LAWs and, in perspective, of all types of weaponized AT
- The only international framework for the discussion of LAWs is the **UN Convention on Certain Conventional Weapons (CCW)** since 2014
- The LAWS have also been discussed in terms of **International Human Rights Law (IHRL)**
- On the International Law applicable to cyber operations , a reference text is the 2017 **Tallinn Manual 2.0**

> *(S. Abaimov and M. Martellini, "AI for the next generation of wars",*
> *PPP for the GP CBRN Sub-working group, Quebec City, 10 October 2018)*

---

[255] Report on Activities, 9-13 April 2018, Campaign to stop killer robots, last retrieved on 8 May 2018 at https://www.stopkillerrobots.org/wp-content/uploads/2018/07/KRC_ReportCCWX_Apr2018_UPLOADED.pdf

[256] A. Conn, European Parliament Passes Resolution Supporting a Ban on Killer Robots 14 September 2018, last retrieved on 18 October 2018 at https://futureoflife.org/2018/09/14/european-parliament-passes-resolution-supporting-a-ban-on-killer-robots/

Those who support LAWS argue that they will be able to meet the requirements of the international laws of war, provide additional military strategic and tactical advantages, add efficiency and spare human emotions and lives of soldiers in dangerous missions.

The Russian government has strongly rejected[257] any ban on lethal autonomous weapons systems, suggesting that such ban could be ignored by adversaries. The same opinion is expressed by the United States, South Korea, and Israel.

The US Army Robotics and Autonomous Systems Strategy[258], published in 2017, states: "Pursuing RAS allows Army Soldiers and teams to defeat capable enemies and maintain overmatch across five capability objectives: increase situational awareness; lighten the warfighters' physical and cognitive workloads; sustain the force with increased distribution, throughput, and efficiency; facilitate movement and manoeuvre; and increase force protection".

The US Department of Defense's "Unmanned Systems Roadmap: 2007-2032"[259] provides additional reasons for pursuing autonomous weapons systems. The additional reasons include that robots are better suited than humans for "'dull, dirty, or dangerous' missions." Examples are presented as follows:

- Dull mission—long-duration sorties;
- Dirty mission—one that exposes humans to potentially harmful CBRNe agents;
- Dangerous mission—explosive ordnance disposal.

In addition, the long-term savings that could be achieved through fielding an army of military robots have been highlighted. In a 2013 article published in The Fiscal Times, David Francis cites Department of Defense figures[260] showing that "each soldier in Afghanistan costs the Pentagon roughly 850,000 USD per year." Some estimate the cost per year to be even higher. Conversely, according to Francis, "the TALON robot—a small rover that can be outfitted with weapons, costs 230,000 USD."

There are also those who advocate to ban not only production and

---

[257] Russia rejects potential UN 'killer robots' ban, official statement says, 1 December 2017, Engineering and Technology, last retrieved on 8 May 2018 at https://eandt.theiet.org/content/articles/2017/12russia-rejects- potential-un-killer-robots-ban-official-statement-says/

[258] The US Army Robotics and Autonomous Systems strategy, March 2017, last retrieved on 8 May 2018 at http://www.arcic.army.mil/App_Documents/RAS_Strategy.pdf

[259] Unmanned Systems Roadmap 2007-2032, Department of Defense, 10 December 2007, last retrieved on 18 June 2018 at https://www.globalsecurity.org/intell/library/reports/2007/dod-unmanned-systems-roadmap_2007-2032.pdf

[260] D. Francis, How a New Army of Robots Can Cut the Defense Budget, 2 April 2013, The Fiscal Times, last retrieved on 18 June 2018 at http://www.thefiscaltimes.com/Articles/2013/04/02/How-a-New-Army-of-Robots-Can-Cut-the-Defense-Budget

deployment, but also research and development. This will be harmful to technological progress, and to the development of machines which should replace human operators in hazardous conditions.

2.9 National Security in Cyberspace

Trying to benefit from all advantages of cyber technologies, but also with full awareness of their threats, nations create cyber security frameworks spreading over and beyond SMART technologies, SMART Cities and infrastructure, e-Mobility, eHealth, eGovernments, etc.

The US Government Accountability Office High-Risk series report (2018)[261] identified four major cybersecurity challenges at the national level:

1. establishing a comprehensive cybersecurity strategy and performing effective oversight,
2. securing federal systems and information,
3. protecting cyber critical infrastructure, and
4. protecting privacy and sensitive data.

The Global Cybersecurity Agenda, global framework for dialogue and international cooperation to coordinate the international response to the growing challenges to cybersecurity and to enhance confidence and security in the information society, launched by ITU in 2007, has seven main strategic goals, built on five work areas: 1) Legal measures; 2) Technical and procedural measures; 3) Organizational structures; 4) Capacity building; and 5) International cooperation, including the elaboration of strategies for the development of model cybercrime legislation.

The seven goals are the following:

1. Elaboration of strategies for the development of a model cybercrime legislation that is globally applicable and interoperable with existing national and regional legislative measures.
2. Elaboration of strategies for the creation of appropriate national and regional organizational structures and policies on cybercrime.
3. Development of a strategy for the establishment of globally accepted minimum security criteria and accreditation schemes for software applications and systems.
4. Development of strategies for the creation of a global framework for watching, warning and incident response to ensure cross-border coordination between new and existing initiatives.
5. Development of strategies for the creation and endorsement of a generic and universal digital identity system and the necessary organizational structures to ensure the recognition of digital credentials for individuals across geographical boundaries.

---

[261] G.L. Dodaro, High-risk series, Urgent Actions are needed to address cybersecurity challenges facing the nation, GAO, last retrieved on 19 August 2018 at https://www.gao.gov/assets/700/693405.pdf

6. Development of a global strategy to facilitate human and institutional capacity-building to enhance knowledge and know-how across sectors and in all the above-mentioned areas.
7. Advice on potential framework for a global multi-stakeholder strategy for international cooperation, dialogue and coordination in all the above-mentioned areas.[262]

Nations identify strategic objectives, priorities and governance frameworks, build capacities for preparedness, response and recovery, find ways of cooperation between the public and private sectors, raise awareness, train, endorse research and development plans, risk assessment plans, ensure incidents monitoring, early warning, inform relevant stakeholders about risks and incidents, etc. National Cyber Security policies are composed based on national information laws, domestic network topology, relations with neighbouring states, available funding, etc. The best practices include a whole government multi-sectoral approach to protect nations from cyber threats vertically and horizontally through joint efforts by political, economic and social forces.

The Global Cybersecurity Index (GCI),[263] a UN multi-stakeholder initiative to measure the commitment of countries to cybersecurity, evaluated their status within five categories: Legal Measures, Technical Measures, Organizational Measures, Capacity Building and Cooperation. The survey, based on the questions related to each of these pillars, was administered by International Telecommunication Unit, a specialized agency of the United Nations, through an online platform with the collected supporting evidence.

The **Global Cybersecurity Index** evaluated the following areas:
1. Legal: Measures based on the existence of legal institutions and frameworks dealing with cybersecurity and cybercrime. Example: cybercriminal legislation, substantive law, procedural cybercriminal law, cybersecurity regulation.
2. Technical: Measures based on the existence of technical institutions and framework dealing cybersecurity. Example: national CIRT, government CIRT, sectoral CIRT, standards for organizations, standardization body.
3. Organizational: Measures based on the existence of policy coordination institutions and strategies for cybersecurity development at the national level. Example: strategy, responsible agency, cybersecurity metrics.
4. Capacity Building: Measures based on the existence of research and development, education and training programmes; certified professionals and public sector agencies fostering capacity building. Example: public awareness, professional

---

[262] Understanding cybercrime: Phenomena, challenges and legal response, ITU, November 2014, last retrieved on 20 August 2018 at https://www.itu.int/en/ITU-D/Cybersecurity/Documents/cybercrime2014.pdf
[263] Global Cybersecurity Index, ITU, last retrieved on 20 August 2018 at https://www.itu.int/en/ITU-D/Cybersecurity/Pages/GCI.aspx

training, national education programmes, R&D programmes, incentive mechanisms, home-grown industry.

5. Cooperation: Measures based on the existence of partnerships, cooperative frameworks and information sharing networks. Example: intra-state cooperation, multilateral agreements, international forums, public-private partnerships, interagency partnerships.

The Global Cybersecurity Index 2017 (GCI 2017) shows that out of the 193 states screened by ITU, there is a big range in cybersecurity commitments.

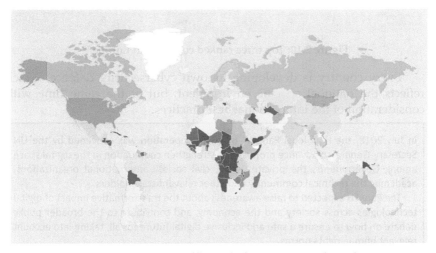

Figure 2.9: GCI Heat Map[264] Level of commitment: from Green (highest) to Red (lowest)

As per the 2017 GCI, the top three ranked countries in Europe are Estonia, France and Norway.

Estonia is the highest-ranking nation in the Europe region. Its enhanced cybersecurity commitment includes the introduction of an organizational structure that can respond quickly to attacks as well as a legal acts that requires all vital services to maintain a minimal level of operation if they are cut off from the Internet. The country also hosts the headquarters of the NATO Cooperative Cyber Defence Centre of Excellence.

France is the second highest ranked in the Europe region, scoring a perfect 100 in capacity building. There is widespread cybersecurity training available in the country, and the National Agency for Information System Security (ANSSI in French) publishes a list of dozens of universities that provide accredited and recognized cybersecurity degrees.

---

[264] Global Cybersecurity Index (GCI) 2017, ITU, 2017, last retrieved on 21 August 2018 at https://www.itu.int/dms_pub/itu-d/opb/str/d-str-gci.01-2017-pdf-e.pdf

Norway is ranked third in Europe with its highest score in the legal pillar. Apart from laws dealing with cybersecurity, Norway has also conducted research on its cybersecurity culture including surveying citizens about the degree to which they will accept monitoring of their online activities.[265]

Country	GCI Score	Legal	Technical	Organizational	Capacity Building	Cooperation
Estonia	0.84	0.99	0.82	0.85	0.94	0.64
France	0.81	0.94	0.96	0.6	1	0.61
Norway	0.78	0.96	0.89	0.64	80.8	0.57

Figure 2.10: Top three ranked countries in Europe[266]

Every country is developing its own cybersecurity defence, which reflects the national specific environment, but at the same time with consideration of the international best practices.

> In July 2018, the High-level Panel on Digital Cooperation was convened by the UN Secretary-General to advance proposals to strengthen cooperation in the digital space among Governments, the private sector, civil society, international organizations, academia, the technical community and other relevant stakeholders.
>
> The Panel is expected to raise awareness about the transformative impact of digital technologies across society and the economy, and contribute to the broader public debate on how to ensure a safe and inclusive digital future for all, taking into account relevant human rights norms.
>
> "The scale, spread and speed of change made possible by digital technologies is unprecedented, but the current means and levels of international cooperation are unequal to the challenge," Mr. Guterres said.
>
> "Digital technologies make a significant contribution to the realization of the Sustainable Development Goals and they cut uniquely across international boundaries. Therefore, cooperation across domains and across borders is critical to realizing the full social and economic potential of digital technologies as well as mitigating the risks that they pose and curtailing any unintended consequences."[267]

The sections below contain examples to illustrate the cyber activities of the selected major cyber powers, i.e. the USA, China and Russia.

---

[265] ibid
[266] Global Cybersecurity Index (GCI) 2017, ITU, 2017, last retrieved on 21 August 2018 at https://www.itu.int/dms_pub/itu-d/opb/str/d-str-gci.01-2017-pdf-e.pdf
[267] Secretary-General's High-level Panel on Digital Cooperation, United Nations Secretary-General, 12 July 2018, last retrieved on 22 August 2018 at https://www.un.org/sg/en/content/sg/personnel-appointments/2018-07-12/secretary-generals-high-level-panel-digital-cooperation

2.9.1 Cyber Security in the USA

Since the 1990s due to increasing cyber-based threats and the persistent nature of information security vulnerabilities, the US designated information security as a government-wide high-risk area. In 2003, it expanded the information security high-risk area to include the protection of critical cyber infrastructure. At that time, the need to manage critical infrastructure protection activities that enhance the security of the cyber and physical public and private infrastructures essential to national security was highlighted, as well as the national economic security, and the national public health and safety. [268]

As per the 2017 GCI, the USA is the top country in the Americas region with cyber commitments (Figure 2.11).

Country	GCI Score	Legal	Technical	Organizational	Capacity Building	Cooperation
United States	0.91	1	0.96	0.92	1	0.73
Canada	0.81	0.94	0.93	0.71	0.82	0.70
Mexico	0.66	0.91	0.89	0.48	0.68	0.34

Figure 2.11: Top three ranked countries in the Americas

The United States of America has the highest scores for the legal and capacity building pillars. The US government has one of the more complex structures of cyber protection systems regulated by the US Department of Homeland Security (DHS), and has been challenged in establishing a comprehensive cybersecurity strategy and in performing effective oversight as called for by federal law and policy. Specifically, it has faced challenges in establishing a comprehensive strategy to provide a framework for how the United States will engage both domestically and internationally on cybersecurity related matters.[269]

Highest Cyber Security Body and Agencies

The US Department of Homeland Security (DHS) is the lead federal department for critical infrastructure protection and non-military federal cybersecurity. It monitors the national access points, controls the US Computer Emergency Response Team, National Incident Response Plan, Critical Infrastructure Protection Plan, and the Einstein program. The

---

[268] GAO, High-Risk Series: An Update, GAO-03-119, Washington, D.C., January 2003
[269] GAO-18-645T High-Risk Series Urgent Actions Are Needed to Address Cybersecurity Challenges Facing the Nation, last retrieved on 17 December 2018 at https://www.gao.gov/products/GAO-18-645T

DHS Cybersecurity Strategy, released in May 2018,[270] sets out five pillars of a DHS-wide risk management approach and provides a framework for executing the cybersecurity responsibilities and leveraging the full range of the Department's capabilities to improve the security and resilience of cyberspace.

The US Department of Homeland Security

To reduce the national cybersecurity risk, DHS applies strategic innovative approaches across the Department and the entire cybersecurity community, engaging with key partners to collectively address cyber vulnerabilities, threats, and consequences. Special attention is paid to reducing and managing the vulnerabilities of federal networks and critical infrastructure; reducing threats from cybercriminal activity through prioritized law enforcement intervention; to mitigating consequences from cybersecurity incidents. It is also committed to engagement with the global cybersecurity community to strengthen the security and resiliency of the overall cyber ecosystems by addressing systemic challenges like increasingly global supply chains; by fostering improvements in international collaboration to deter malicious cyber actors and build capacity; by increasing research and development, and by improving cyber workforce.[271]

Other government agencies, e.g., Department of Defense (DoD), Department of Energy, Department of State, etc., have their own security infrastructure and companies maintain their own networks in compliance with regulations such as the National Institute of Standards & Technology (NIST) 800-XX series, the Federal Information Security Management Act (FISMA) of 2002, or Homeland Security/Presidential Directives.

The FBI is the lead federal agency for investigating cyberattacks by criminals, overseas adversaries, and terrorists.[272] And the other cyber-related organizations are the Federal Aviation Administration (FAA) with the Next Generation Air Transportation System (NextGen) project for modernization of the National Airspace System (NAS); the Department of Energy with the Smart Grid, the individual services (Army, Air Force, Navy, and Marines) which have the authority and budget to decide how to implement cybersecurity. Each branch of the service has a name for their portion of the network. Defense Information Systems Agency (DISA) runs the Global Information Grid (GIG),[273] the DoD IT infrastructure,

---

[270] DHS Cybersecurity Strategy, Homeland Security, last retrieved on 18 December 2018 at https://www.dhs.gov/publication/dhs-cybersecurity-strategy

[271] Cybersecurity, Homeland Security, last retrieved on 18 December 2018 at https://www.dhs.gov/topic/cybersecurity

[272] Cyber Crime, FBI, last retrieved on 17 September 2018 at https://www.fbi.gov/investigate/cyber

[273] Global Information Grid, JROCM 134-01, 30 August 2001, last retrieved on 17 December 2018 at http://www.acqnotes.com/Attachments/Global%20Information%20Grid%20Capstone%20Requirements%20Document,%2030%20Aug%2001.pdf

Air Force has C2 Constellation, the Army has LandWarNet, and Navy has FORCEnet, components of GIG.

Since 2009, the National Cybersecurity and Communications Integration Center (NCCIC) has served as the national hub for cyber and communications information, technical expertise, and operational integration. The US Computer Emergency Response Team (US-CERT), which is one of its branches, is a central Federal information security incident centre that compiles and analyses information about incidents that threaten information security. Federal agencies are required to report such incidents to US-CERT.[274]

In 2009, the US formed the United States Cyber Command (USCYBERCOM) under the US Strategic Command (USSTRATCOM), which has a mission to "ensure US freedom of action in space and cyberspace". It is composed of units from the Army Forces Cyber Command, the Fleet Cyber Command, the 24th Air Force, and the Marine Corps Forces Cyberspace Command. In August 2017, USCYBERCOM was elevated from a sub-unified command to a Unified Combatant Command responsible for cyberspace operations, which was seen as recognition of the growing centrality of cyberspace to the US national security.[275]

The National Cyber Security Division (NCSD) is a division which operates under the Directorate for National Protection and Programs (DHS). It collaborates with the private sector, government, military, and intelligence stakeholders to conduct risk assessments and mitigate vulnerabilities and threats to information technology assets and activities affecting the operation of the civilian government and private sector critical cyber infrastructure. NCSD also provides cyber threat and vulnerability analysis, early warning, and incident response assistance for public and private sector constituents. NCSD carries out the majority of DHS' responsibilities under the Comprehensive National Cybersecurity Initiative.

Regulations

The three basic federal regulations mandate healthcare organizations, financial institutions, and federal agencies to protect their systems and information.

- 1996 Health Insurance Portability and Accountability Act (HIPAA).
- 1999 Gramm-Leach-Bliley Act.
- 2002 Homeland Security Act, which included the Federal Information Security Management Act (FISMA).

---

[274] About US, US-CERT, last retrieved on 15 September 2018 at https://www.us-cert.gov/about-us

[275] About Us, U.S. Cyber Command, last retrieved on 17 September 2018 at https://www.cybercom.mil/About/

While the term "cybersecurity" is mostly associated with the Western reality, it is worth noting that the 1998 US "Joint Doctrine for Information Operations" described how joint forces use information operations (IO) to support the US national military strategy. It discussed integration and synchronization of offensive and defensive IO in the planning and execution of combatant commanders' plans and operations to support the strategic, operational, and tactical levels of war.[276] Doctrine stated that "IO capitalize on the growing sophistication, connectivity, and reliance on information technology. IO target information or information systems in order to affect the information-based process, whether human or automated. Such information dependent processes range from National Command Authorities-level decision making to the automated control of key commercial infrastructures such as telecommunications and electric power... The thoughtful design and correct operation of information systems are fundamental to the overall conduct of IO."[277]

However, at a certain point the new term "cybersecurity" evolved and among the further regulations, there are the Cybersecurity Enhancement Act of 2014, Cybersecurity Information Sharing Act (CISA) of 2015, Federal Exchange Data Breach Notification Act of 2015, National Cybersecurity Protection Advancement Act of 2015 (which amended the Homeland Security Act of 2002). The US National Counterintelligence Strategy of 2016 was developed in accordance with the Counterintelligence Enhancement Act of 2002. The Strategy sets forth how the US will identify, detect, exploit, disrupt, and neutralize foreign intelligence entity (FIE) threats. It provides guidance for the counterintelligence programs and activities of the U.S. Government intended to mitigate such threats, including in cyberspace.

Table 2.4: Recent executive branch initiatives that identify cybersecurity priorities for the Federal Government Executive branch initiative[278]

Initiatives	Date of issuance	Description
Executive Order 13800: Strengthening the Cybersecurity of Federal Networks and Critical Infrastructure	May 2017	The Executive Order required federal agencies to take a variety of actions, including better management of their cybersecurity risks and to coordinate meeting and reporting requirements related

(Contd.)

---

[276] Joint Doctrine for Information Operations, 9 October 1998, last retrieved on 17 September 2018 at http://www.c4i.org/jp3_13.pdf
[277] Ibid
[278] GAO-18-645T High-Risk Series Urgent Actions Are Needed to Address Cybersecurity Challenges Facing the Nation, last retrieved on 17 September 2018 at https://www.gao.gov/products/GAO-18-645T

Table 2.4: *(Contd.)*

		to the cybersecurity of federal networks, critical infrastructure, and the nation.[279] As of July 2018, the executive branch had publicly released several reports, including a high-level assessment by the Office of Management and Budget (OMB) of the cybersecurity risk management capabilities of the federal government.[280] The assessment stated that OMB and the Department of Homeland Security (DHS) examined the capabilities of 96 civilian agencies across 76 cybersecurity metrics and found that 71 agencies had cybersecurity programs that were either at risk or high risk.[281] The report also stated agencies were not equipped to determine how malicious actors seek to gain access to their information systems and data. The report identified core actions to address cybersecurity risks across the federal enterprise.
National Security Strategy	December 2017	The National Security Strategy[282] identified four vital national interests: protecting the homeland, the American people, and American way of life; promoting American prosperity; preserving peace through strength; and advance American influence. The strategy also cites cybersecurity as a national priority and identifies related needed actions, including

(Contd.)

---

[279] Presidential Executive Order on Strengthening the Cybersecurity of Federal Networks and Critical Infrastructure. Executive Order 13800 (Washington, D.C.: May 11, 2017).

[280] OMB, Federal Cybersecurity Risk Determination Report and Action Plan (Washington, D.C.: May 2018).

[281] OMB and DHS designated agencies as "at risk" if agencies had some essential policies, processes, and tools in place to mitigate overall cybersecurity risks. OMB and DHS designated agencies as "high risk" if agencies did not have essential policies, processes, and tools in place to mitigate overall cybersecurity risks.

[282] The President of the United States, National Security Strategy of the United States of America (Washington, D.C.: Dec. 2017).

Table 2.4: (*Contd.*)

Initiatives	Date of issuance	Description
DHS Cybersecurity Strategy	May 2018	identifying and prioritizing risk, building defensible government networks, determining and disrupting malicious cyber actors, improving information sharing and deploying layered defences. The DHS Cybersecurity Strategy[283] articulated seven goals the department plans to accomplish in support of its mission related to managing national cybersecurity risks. The goals were spread across five pillars that correspond to DHS-wide risk management, including risk identification, vulnerability reduction, threat reduction, consequence mitigation, and enabling cybersecurity outcomes. The strategy is intended to provide DHS with a framework to execute its cybersecurity responsibilities during the next 5 years to keep pace with the evolving cyber risk landscape by reducing vulnerabilities and building resilience; countering malicious actors in cyberspace; responding to incidents; and making the cyber ecosystem more secure and resilient.

Budgets

The US funding for Research and Development activities in the Department of Defense for FY19 was considerably increased with strong emphasis on technology development and prototyping. The increase included a half-billion-dollar or 19 percent increase in basic research.[284] The FY2019

---

[283] DHS, US Department of Homeland Security Cybersecurity Strategy (Washington, D.C.: May 2018).

[284] FY19 Appropriations Bills: Department of Defense S&T, 19 July 2018, last retrieved on 18 September 2018 at https://www.aip.org/fyi/2018/fy19-appropriations-bills-department-defense-st?utm_medium=email&utm_source=FYI&dm_i=1ZJN,5RH06,TA6J5U,MGRBW,1

measure provides 607.1 billion USD in the base of Department of Defense funding and 67.9 billion USD in OCO funding.[285]

The Senate Appropriations Committee highlighted its proposals for funding in high-priority technology areas in excess of the administration's budget request. These include an additional 929 million USD for offensive and defensive technologies related to hypersonic propulsion, an additional 564 million USD for space capabilities, an additional 447 million USD for trusted microelectronics and chip manufacturing, an additional 317 million USD for directed energy technologies, and an additional 308 million USD for artificial intelligence.

According to the budget documents, in FY18, 5.4 million USD of overseas contingency operations funds will be spent toward JTF-Ares. In another Air Force line item, base budget and OCO funds totalling 80 million USD will be spent on cyber tools.[286]

The OCO funding will go specifically toward JTF-A while base funds will provide a wider array of services. These include:

- Development of deployed exploitation framework for Cyber Command.
- Execution of a spiral development process for cyberspace operations basic tools to provide operational agility during cyber mission force effects operations.
- Continued tool repository and a signature management study on each spiral of delivered tools that enables tool measurement and repository as well as a means to manipulate tool code to minimize risk of discovery.

Under a separate line item, the Air Force will spend 35 million USD of the base budget, 4 million USD OCO for a total of 39 million USD for offensive cyberspace operations. Meanwhile the Army, for its part, under a research, development test and evaluation line will spend 7 million USD in FY18 on a line item titled "Offensive Information Operations Technologies."

Under House and Senate spending proposals for the fiscal year 2019, the DoD Research, Development, Test, and Evaluation budget is set to continue its recent ascent. Senate appropriators are recommending an 8

---

[285] FY2019 Defense Appropriations Bill Gains Senate Committee Approval, United States Senate Committee on Appropriations, 28 June 2018, last retrieved on 18 September 2018 at https://www.appropriations.senate.gov/news/fy2019-defense-appropriations-bill-gains-senate-committee-approval

[286] M. Pomerleau, What are services spending on offensive cyber? 24 May 2017, last retrieved on 17 September 2018 at https://www.fifthdomain.com/2017/05/24/what-are-services-spending-on-offensive-cyber/

percent increase from 88.3 billion to 95.1 billion USD, while the House proposes a more modest 3 percent increase to 91.2 billion USD. The RDT&E budget stood at 72.3 billion USD in fiscal year 2017. This increase reflects Congress and DoD's determination that the US needs to invest heavily in cutting-edge military technologies. The new funding is flowing predominantly to later-stage development and testing activities to help accelerate the transition of technologies into acquisition programs.

The US FY19 National Defense Authorization Act budget amounts to 717 billion USD,[287] and includes 15 million USD to "enhance and accelerate artificial intelligence research" in the service branches—10 million USD for the Air Force and 5 million USD for the Army.

The controversial Project Maven, the Defense Department's program to use machine learning to identify and categorize objects in drone imagery, received a 580 percent funding increase in this year's bill—from 16 million USD in 2018 to 93.1 million USD in 2019. As AI and machine learning algorithms are integrated into defence tech, spending is only going to increase in years to come.[288]

The National Defense Authorization Act (NDAA) also authorized 10 million USD to be spent on a new National Security Commission on Artificial Intelligence. The commission will have 15 members, appointed by various offices, including the Secretaries of Defense and Commerce, and the Chairs and Ranking Members of the House and Senate Armed Services Committees. It will be tasked with assessing how the Defense Department can better use artificial intelligence and machine learning capabilities for national security, as well as the associated national security risks and ethical considerations related to them.

Plans of Action

The Cybersecurity Strategy and Implementation Plan for the Federal Civilian Government was issued in October 2015.[289] The plan directed a series of actions to improve capabilities for identifying and detecting vulnerabilities and threats, enhance protections of government assets and information, and further develop robust response and recovery capabilities to ensure readiness and resilience when these incidents inevitably occur. The plan also identified key milestones for major

---

[287] H.R.5515 - John S. McCain National Defense Authorization Act for Fiscal Year 2019, 115th Congress (2017-2018), last retrieved on 17 September 2018 at https://www.congress.gov/bill/115th-congress/house-bill/5515/text#toc-H6C2FA09C23154F80B0D293929D5ACFB5

[288] Pentagon's artificial intelligence programs get huge boost in defense budget, 15 August 2018, last retrieved on 18 August 2018 at https://www.fastcompany.com/90219751/pentagons-artificial-intelligence-programs-get-huge-boost-in-defense-budget

[289] The National Cyber Incident Response Plan (NCIRP), US-CERT, last retrieved on 18 August 2018 at https://www.us-cert.gov/ncirp

activities, resources needed to accomplish these milestones, and the specific roles and responsibilities of federal organizations related to the strategy's milestones.

The National Cyber Incident Response Plan (NCIRP),[290] released by DHS in December 2016, outlines a national approach to cyber incidents; addresses the important role that the private sector, state and local governments, and multiple federal agencies play in responding to incidents and how the actions of all of these fit together for an integrated response; it reflects and incorporates lessons learned from exercises, real world incidents and policy and statutory updates, such as the Presidential Policy Directive/PPD-41: U.S. Cyber Incident Coordination, and the National Cybersecurity Protection Act of 2014. The NCIRP also serves as the Cyber Annex to the Federal Interagency Operational Plan (FIOP) that built upon the National Planning Frameworks and the National Preparedness System.

National Programmes and Research

Two following research programmes, among others, support the US government in cyber security.

NSA and DHS jointly sponsor the National Centers of Academic Excellence in Information Assurance (IA) Education and research programmes. The goal of these programmes is to reduce vulnerability in the national information infrastructure by promoting higher education and research in IA and producing a growing number of professionals with IA expertise in various disciplines.[291]

The Defense Advanced Research Projects Agency (DARPA) is a DOD agency responsible for the development of emerging technologies for use by the military. Currently, DARPA's mission statement is "to make pivotal investments in breakthrough technologies for national security."[292]

Among the current DARPA programmes there are the Active Social Engineering Defense programs aimed to develop core technology to enable the capability to automatically elicit information from a malicious adversary in order to identify, disrupt, and investigate social engineering attacks; the Software Defined Hardware program to build runtime-reconfigurable hardware and software that enables near application-specific integrated circuit performance without sacrificing programmability for data-intensive algorithms; the Domain-specific System on Chip (SoC)

---

[290] National Cyber Incident Response Plan, Homeland Security, December 2016, last retrieved on 28 August 2018 at https://www.hsdl.org/?view&did=798128

[291] National Centers of Academic Excellence, Central Security Service, NSA, last retrieved on 28 August 2018 at https://www.nsa.gov/resources/students-educators/centers-academic-excellence/

[292] Mission, Defense Advanced Research Projects Agency, last retrieved on 28 August 2018 at https://www.darpa.mil/about-us/mission

program is to develop a heterogeneous SoC comprised of many cores that mixes general-purpose processors, special-purpose processors, hardware accelerators, memory, and input/output. DSSoC seeks to enable the rapid development of multi-application systems through a single programmable device; Intelligent Design of Electronic Assets to create a "no human in the loop" layout generator that enables users with no electronic design expertise to complete the physical design of electronic hardware within 24 hours; Posh Open Source Hardware programs to create an open source SoC design and verification ecosystem that will enable the cost effective design of ultra-complex SoCs, etc.[293]

Private companies invest larger resources into cutting-edge technologies, employing experts, offering multimillion-dollar pay packages that the government can never match.

One of the top research and development centres in the world is the Silicon Valley where thousands of technology companies are headquartered. It has the largest concentration of high-tech companies in the United States, at 387,000 high-tech jobs, of which Silicon Valley accounts for 225,300 high-tech jobs. Silicon Valley has the highest concentration of high-tech workers of any metropolitan area, with 285.9 out of every 1,000 private-sector workers.

Military and Intelligence communities have always been collaborating with Silicon Valley companies. For example, David Packard, Hewlett-Packard's co-founder, served as the Deputy Secretary of Defence under President Richard M. Nixon. Though there is a growing antagonism between the government agencies and private companies. "In the wake of Edward Snowden, there has been a lot of concern over what it would mean for Silicon Valley companies to work with the national security community," said Gregory Allen[294], an adjunct fellow with the Center for a New American Security. "These companies are—understandably—very cautious about these relationships."

> The Deputy Defense Secretary Bob Work announced in a memo that he was establishing an Algorithmic Warfare Cross-Functional Team, overseen by the Undersecretary of Defense for intelligence, to work on something he called Project Maven.
>
> "As numerous studies have made clear, the Department of Defense must integrate artificial intelligence and machine learning more effectively across operations to maintain advantages over increasingly capable adversaries and competitors," Work wrote.
>
> "Although we have taken tentative steps to explore the potential of artificial intelligence, big data and deep learning," he added, "I remain convinced that we need

---

[293] Offices, Defense Advanced Research Projects Agency, last retrieved on 28 August 2018 at https://www.darpa.mil/about-us/offices

[294] Pentagon wants Silicon Valley's help on A.I., The New York Times, 15 March 2018, last retrieved on 1 August 2018 at https://www.nytimes.com/2018/03/15/technology/military-artificial-intelligence.html

to do much more and move much faster across DoD to take advantage of recent and future advances in these critical areas."

Project Maven focuses on computer vision—an aspect of machine learning and deep learning—that autonomously extracts objects of interest from moving or still imagery, Cukor said. Biologically inspired neural networks are used in this process, and deep learning is defined as applying such neural networks to learning tasks.

"This effort is an announcement . . . that we're going to invest for real here," he said.[295]

The Project Maven new task force includes Terah Lyons, the executive director of the Partnership on AI, an industry group that includes many of Silicon Valley's biggest companies.

"It gives me great pleasure to announce that the US Undersecretary of Defense for Intelligence—USD(I)—has awarded Google and our partners a contract for 28 million USD, 15 million USD of which is for Google ASI, GCP, and PSO," Scott Frohman, a defense and intelligence sales lead at Google, wrote in a September 29, 2017 email. "Maven is a large government program that will result in improved safety for citizens and nations through faster identification of evils such as violent extremist activities and human right abuses. The scale and magic of Google Cloud Platform, the power of Google ML [machine learning], and the wisdom and strength of our people will bring about multi-order-of-magnitude improvements in safety and security for the world."[296]

Google's decision to provide artificial intelligence to DoD for the analysis of drone footage has led to a protest within Google, and some employees and academics have resigned because of the company's involvement in Project Maven. Many of them argued that humans should be tasked with categorizing drone images rather than computers, since those inferences could be used to determine lethal drone strikes. Google will not renew its current contract to provide artificial intelligence to the US Department of Defense for analysing drone footage after its current contract expires in 2019.

In July 2018, Booz Allen Hamilton, a consulting firm, signed a 5-year 885 million USD artificial intelligence contract with the US Government Program Office. It will focus on integrating machine learning capabilities with a full spectrum of Command, Control, Communications, Computers, Intelligence, Surveillance and Reconnaissance (C4ISR) systems that support military operations.[297]

---

[295] Pentagon wants Silicon Valley's help on A.I., The New York Times, 15 March 2018, last retrieved on 1 August 2018 at https://www.nytimes.com/2018/03/15/technology/military-artificial-intelligence.html

[296] K. Cogner, Google Plans Not to Renew Its Contract for Project Maven, a Controversial Pentagon Drone AI Imaging Program, 1 June 2018, last retrieved on 28 August 2018 at https://gizmodo.com/google-plans-not-to-renew-its-contract-for-project-mave-1826488620

[297] Press Release, U.S. Government & GSA FEDSIM Select Booz Allen to Help Apply Artificial Intelligence, 30 July 2018, last retrieved on 17 September 2018 at https://www.boozallen.com/e/media/press-release/booz-allen-selected-to-help-apply-artificial-intelligence.html

Nuclear Weapons

As stated in the Nuclear Posture Review,[298] in February 2018, the United States has been strongly supporting the reduction of the number of nuclear weapons. The 1991 Strategic Arms Reduction Treaty (START) set a ceiling of 6,000 accountable strategic nuclear warheads—a deep reduction from Cold War highs. Shorter-range nuclear weapons were almost entirely eliminated from America's nuclear arsenal in the early 1990s. The 2002 Strategic Offensive Reduction Treaty and the 2010 New START Treaty further lowered strategic nuclear force levels to 1,550 accountable warheads. During this time, the U.S. nuclear weapons stockpile came down by more than 85 percent from its Cold War high. Many hoped conditions had been set for even deeper reductions in global nuclear arsenals, and, ultimately, for their elimination.

The United States will continue its efforts to: 1) minimize the number of nuclear weapons states, including by maintaining credible U.S. extended nuclear deterrence and assurance; 2) deny terrorist organizations access to nuclear weapons and materials; 3) strictly control weapons-usable material, related technology, and expertise; and 4) seek arms control agreements that enhance security, and are verifiable and enforceable.[299]

Cyber Events

A big number of information security incidents were reported by federal executive branch civilian agencies to DHS's US Computer Emergency Readiness Team (US-CERT). For the fiscal year 2017, 35,277 such incidents were reported by the Office of Management and Budget (OMB) in its 2018 annual report to Congress, as mandated by FISMA.[300] Among these there are web-based attacks, phishing, and the loss or theft of computing equipment.

The US Government Accountability Office High Risk series report issued in July 2018 indicates the following Federal Information Security Incidents for FY 2017.[301]

- In March 2018, the Mayor of Atlanta, Georgia reported that the city was victimized by a ransomware cyberattack. As a result, city government officials stated that customers were not able to access multiple applications that are used to pay bills or access court related

---

[298] Nuclear posture review, February 2018, Office of the Secretary of Defence, 2 February 2018, last retrieved on 17 September 2018 at https://media.defense.gov/2018/Feb/02/2001872886/-1/-1/1/2018-NUCLEAR-POSTURE-REVIEW-FINAL-REPORT.PDF

[299] Ibid

[300] The Federal Information Security Modernization Act of 2014 was enacted as Pub. L. No. 113-283, 128 Stat. 3073 (Dec. 18, 2014), and amended chapter 35 of Title 44, US Code.

[301] GAO Report: High Risk series report, Urgent actions are needed to address cybersecurity challenges facing the nation, GAO-18-645T, 25 July 2018

information. In response to the attack, the officials noted that they were working with numerous private and governmental partners, including DHS, to assess what occurred and determine how best to protect the city from future attacks.

- In March 2018, the Department of Justice reported that it had indicted nine Iranians for conducting a massive cybersecurity theft campaign on behalf of the Islamic Revolutionary Guard Corps. According to the department, the nine Iranians allegedly stole more than 31 terabytes of documents and data from more than 140 American universities, 30 US companies, and five federal government agencies, among other entities.
- In March 2018, a joint alert from DHS and the Federal Bureau of Investigation 16 stated that, since at least March 2016, Russian government actors had targeted the systems of multiple US government entities and critical infrastructure sectors. Specifically, the alert stated that Russian government actors had affected multiple organizations in the energy, nuclear, water, aviation, construction, and critical manufacturing sectors.
- In July 2017, a breach at Equifax resulted in the loss of PII for an estimated 148 million US consumers. According to Equifax, hackers accessed people's names, Social Security numbers, birth dates, addresses and, in some instances, driver's license numbers.
- In April 2017, the Commissioner of the Internal Revenue Service testified that the IRS had disabled its data retrieval tool in early March 2017 after becoming concerned about the misuse of taxpayer data. Specifically, the agency suspected that PII obtained outside the agency's tax system was used to access the agency's online federal student aid application in an attempt to secure tax information through the data retrieval tool. In April 2017, the agency began notifying taxpayers who could have been affected by the breach.
- In August 2015, the Office of Personnel Management of the U.S. government revealed that approximately 22 million personnel records of U.S. government employees—including those with high-level security clearances—had been compromised. These records contained information that went far beyond basic identifying information and, in the case of those who had applied for security clearances, included fingerprints and lists of foreign contacts.[302]
- In June 2015, OPM reported that an intrusion into its systems had affected the personnel records of about 4.2 million current and former federal employees. Then, in July 2015, the agency reported that a

---

[302] Mike Levine and Jack Date, "22 Million Affected by OPM Hack, Officials Say," ABC News, 9 July 2015, last retrieved on 17 September 2018 at http://abcnews.go.com/US/exclusive-25-million-affected-opm-hack-sources/story?id=32332731

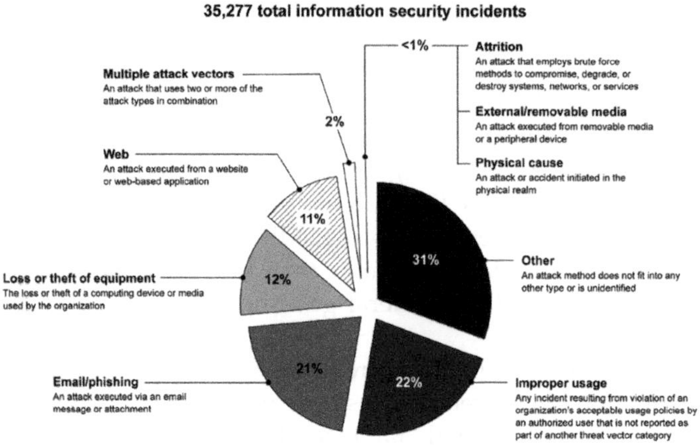

35,277 total information security incidents

Figure 2.12: Federal Information Security Incidents by Threat
Vector Category, FY 2017
Source: GAO High Risk series report, July 2018

safeguarding federal IT system and the systems that support critical
infrastructures has been a long-standing concern of GAO.

- In March 2015, Premera Blue Cross, a health insurer based in
  Washington State, reported that the personal information of up to
  11 million customers could have been exposed in a data breach that
  occurred in 2014.[303]
- In February 2015, Anthem, one of the largest health insurers in the
  United States, announced that it had been the target of an effort to
  obtain the personal information of tens of millions of its customers
  and employees. The information in question included names,
  Social Security numbers, birthdays, addresses, email addresses, and
  employment information, including income data.[304]
- In September 2014, Home Depot reported that about 56 million credit
  and debit cards had probably been compromised over a six-month
  period earlier that year through malicious software implanted on
  point-of-sale terminals.[305]

---

[303] A. Vijayan, "Premera Hack: What Criminals Can Do with Your Healthcare
Data," Christian Science Monitor, 20 March 2015, last retrieved on 17
September 2018 at http://www.csmonitor.com/World/Passcode/2015/0320/
Premera- hack-What-criminals-can-do-with-your-healthcare-data.

[304] R. Abelson and M. Goldstein, "Millions of Anthem Customers Targeted in Cyberattack,"
New York Times, 5 February 2015, last retrieved on 19 September 2018 at http://www.
nytimes.com/2015/02/05/business/hackers-breached-data-of-millions-insurer-says.
html.

[305] Julie Creswell and Nicole Perlroth, "Ex-Employees Say Home Depot Left Data
Vulnerable," New York Times, 19 September 2014, last retrieved on 19 September 2018
at http://www.nytimes.com/2014/09/20/business/ex-employees-say-home-depot-left-
data-vulnerable.html.

- In May 2014 the U.S. Justice Department issued indictments against five members of the Chinese People's Liberation Army for violations of the Computer Fraud and Abuse Act (CFAA) and the Economic Espionage Act, alleging that these individuals engaged in criminal acts of industrial espionage that took place in the 2006–2014 period.[306]
- In June 2013 the U.S. Department of Defense acknowledged that sensitive unclassified data regarding the F-35 fighter jet had been stolen, significantly reducing the U.S. design and production edge on fifth-generation fighters (e.g., cost advantage and lead time) compared to other nations that are seeking to produce such fighters.[307]

2.9.2 Cyber Security in Russia

The GCI 2017 Report ranked Russia as second in the Commonwealth of Independent States.[308]

Country	GCI Score	Legal	Technical	Organizational	Capacity Building	Cooperation
Georgia	0.81	0.91	0.77	0.82	0.9	0.7
Russian Federation	0.78	0.82	0.67	0.85	0.91	0.7
Belarus	0.59	0.85	0.63	0.33	0.68	0.47

Figure 2.13: Top three ranked countries in Commonwealth of Independent States
Source: Global Cybersecurity Index (GCI) 2017, ITU, 2017

The Russian Federation, ranked second in the region, scores best in capacity building. Its commitments range from developing cybersecurity standards to R&D and from public awareness to a home-grown cybersecurity industry. An example of the latter is Kaspersky Labs, founded in 1997 and whose software protects over 400 million users and some 270,000 organizations.[309]

As it was discussed in the previous chapters, Russia uses the concept of "information security", which has a broader meaning and covers both the security of the information content and the devices which store, process and transmit information. Consequently, Information-technological

---

[306] Department of Justice, Office of Public Affairs, "U.S. Charges Five Chinese Military Hackers for Cyber Espionage against U.S. Corporations and a Labor Organization for Commercial Advantage," 19 May 2014, last retrieved on 19 September 2018 at http://www.justice.gov/opa/pr/us-charges-five-chinese-military-hackers-cyber-espionage-against-us-corporations-and-labor.

[307] David Alexander, "Theft of F-35 Design Data Is Helping U.S. Adversaries—Pentagon," Reuters, 19 June 2013, last retrieved on 19 September 2018 at http://www.reuters.com/article/2013/06/19/usa-fighter-hacking-idUSL2N0EV0T320130619.

[308] Global Cybersecurity Index (GCI) 2017, ITU, 2017, last retrieved on 18 October 2018 at https://www.itu.int/dms_pub/itu-d/opb/str/d-str-gci.01-2017-pdf-e.pdf

[309] Kaspersky Labs, Official Website, last retrieved on 19 September 2018 at https://www.kaspersky.com

attacks are described as cyberattacks ranging from influence operations via website defacements to physical damage resulting from altered rocket trajectories. Information-psychological operations, in contrast, include operations that attack the morale and the perceptions of the population, as exemplified by the Arab Spring, Orange Revolution, and even the dissolution of the Soviet Union.[310]

Russian industry plays an important role in the Russian cyber security: telecommunications infrastructure, hardware supply chain, and information security technology sector. Russia's telecommunication sector, which is the foundation of the national Internet communication, broadband and mobile communications sectors are highly consolidated; just six companies dominate 77.1% of the broadband market, while 92% of the mobile market is controlled by four operators.[311] Although most of these companies are privately owned, the largest operator that controls one third of all of Russian broadband access, Rostelecom, is state-owned.[312] These companies operate under the direct supervision of the Russian government. Additionally, some of these companies operate abroad and effectively rely on Russian cyber terrain to access their own cyber resources.[313]

Cyber Security Bodies and Agencies

The President of the Russian Federation is the head of the states bodies for information security and directs the Security Council and approves information security decrees. The general management of the information security system is carried out by the President and the Government of the Russian Federation.

The Security Council under the President of the Russian Federation is the Executive body directly responsible for state security issues. This includes the Interagency Commission on Information Security which prepares presidential decrees, acts as a legislative initiative, coordinates the activities of heads of ministries and departments in the field of information security of the state. Its working body is the State Technical Commission under the President of the Russian Federation responsible for certification of information security tools (except cryptographic tools), licensing of activities in the field of production of security tools

---

[310] I. Sheremet, "Kiberugrozy Rossii Rastut—Chast' II [Cyberthreats to Russia Grow - Part II," Voyenno-promyshlennyy Kur'yer, 19 February 2014, last retrieved on 20 December 2018 at http://vpk-news.ru/articles/19194

[311] S. Kelly et al., Freedom on the Net 2014: Russia, Freedom on the Net (Freedom House, 2014): 3, last retrieved on 19 September 2018 at https://freedomhouse.org/sites/default/files/resources/Russia.pdf.

[312] Ibid.

[313] P. Tucker, "Why Ukraine Has Already Lost the Cyberwar, Too," Defense One, 28 April 2014, last retrieved on 10 October 2018 at http://www.defenseone.com/technology/2014/04/why-ukraine-has-already-lost-cyberwar-too/83350/

and training of information security specialists. It also coordinates and directs the activities of state research institutions working in the field of information security, provides accreditation of licensing bodies and testing centres (laboratories) for certification.

The Interagency Commission for the protection of state secrets is charged with the task of managing the licensing of enterprises, institutions and organizations associated with the use of information constituting a state secret, with the creation of means of protection of information, as well as the provision of services for the protection of state secrets. In addition, the Commission coordinates the certification of information security.

The Federal Agency for Government Communications and Information under the President of the Russian Federation (FAPSI) provides government communications and information technologies for the public administration. The Agency certifies all means used for the organization of government communications and information management of public administration, as well as licenses for all enterprises, institutions and organizations engaged in the production of such means. In addition, the FAPSI is exclusively responsible for certification and licensing in the field of cryptographic protection of information.

Supervision of compliance with the law on Mass Media and Mass Communications, Television and Radio Broadcasting is implemented by the Federal Service for Supervision of Communications and Information Technology (Roskomnadzor) who is also responsible for supervision of the general compliance with personal data laws.[314] This agency is also enabled to blacklist and shutdown non-compliant websites without trial. Thus, 62,000 web sites or web pages with illegal information were blocked in 2017. Illegal information was voluntarily removed by owners from 112 thousand sites or web pages after notifications of Roskomnadzor.[315] The blogging sites that attract a daily audience of more than 3,000 visitors must register with Roskomnadzor, with bloggers required to provide personally identifiable information and assume financial liability for the accuracy of the contents on their blogs.

The information security framework of the Russian Federation also includes the Federal Information Service, Foreign Intelligence Service of the Russian Federation, Ministry of Internal Affairs of the Russian Federation, Judiciary bodies, State Customs Committee of the Russian Federation., State Standard of the Russian Federation (Gosstandart), etc.

---

[314] Resolution of 16 March 2009 No 228 on Roskomnadzor, Federal service for supervision of communications and information technology of the Russian Federation, last retrieved on 20 December 2018 at https://rkn.gov.ru/about/p179/

[315] What and how is blocked by Roskomnadzor, 12 March 2018, last retrieved on 20 December 2018 at http://d-russia.ru/chto-i-kak-blokiruet-roskomnadzor.html

Cyber Security Regulations

The beginning of 2000s saw a sequence of security regulations developed by the Russian Federation.

The Military Doctrine of the Russian Federation (Military Constitution), the first version adopted in 2000 and the latest one was approved in December 2014,[316] is one of the main documents of strategic planning in the Russian Federation. It considers the main provisions of the Concept of long-term socio-economic development of the Russian Federation for the period up to 2020, the Strategy of the National Security of the Russian Federation up to 2020, as well as the relevant provisions of the Concept of the Foreign Policy of the Russian Federation and the Maritime doctrine of the Russian Federation for the period up to 2020, etc.

Article 15 of the Doctrine enumerates the characteristics and features of modern military conflicts highlighting among others massive use of weapons systems and military equipment, high-precision, hypersonic weapons, electronic warfare, weapons on new physical principles, comparable in efficiency with nuclear weapons, information and control systems, as well as unmanned aerial and autonomous naval vehicles, controlled robotic weapons and military equipment.

In December 2015, the Decree of the President of the Russian Federation No. 683 approved the new National Security Strategy of the Russian Federation, "the basic document for strategic planning, defining national interests and national strategic priorities of the Russian Federation, as well as the goals and measures, both domestically and in foreign policy, that are meant to strengthen national security … and to ensure the sustainable development of the country in a long-term perspective."[317]

The Doctrine of Information Security of the Russian Federation, initially adopted in 2000 and in its latest version from 5 December 2016, defines strategic objectives and key areas of information security taking into account the strategic national priorities of the Russian Federation. It is a strategic planning document on ensuring the national security of the Russian Federation, which builds upon the provisions of the National Security Strategy of the Russian Federation, and other strategic planning documents in this area. The Doctrine provides the basis for

---

[316] Military Doctrine of the Russian Federation, adopted on 5 February 2014, last retrieved on 21 December 2018 at http://www.mid.ru/documents/10180/822714/41d527556bec8deb3530.pdf/d899528d-4f07-4145-b565-f

[317] National Security Strategy of the Russian Federation of 31 December 2015, Order of the President of the Russian Federation No 683, last retrieved on 19 December 2018 at https://rg.ru/2015/12/31/nac-bezopasnost-site-dok.html

the development of the State policy and public relations in the sphere of information security.[318]

Doctrine of Information Security of the Russian Federation[319]
The Doctrine uses the following basic notions:

(a) the national interests of the Russian Federation in the information sphere (hereinafter referred to as the "national interests in the information sphere") are the objectively meaningful needs of the individual, society and the State in ensuring their safety and security and sustainable development in the information sphere;

(b) the threat to the information security of the Russian Federation (hereinafter referred to as the "information threat") is a combination of actions and factors creating a risk of damaging the national interests in the information sphere;

(c) the information security of the Russian Federation (hereinafter referred to as the "information security") is the state of protection of the individual, society and the State against internal and external information threats, allowing them to ensure the constitutional human and civil rights and freedoms, the decent quality and standard of living for citizens, the sovereignty, the territorial integrity and sustainable socio-economic development of the Russian Federation, as well as defence and security of the State;

(d) the provision of information security is the implementation of mutually supportive measures (legal, organizational, investigative, intelligence, counter-intelligence, scientific and technological, information and analytical, personnel-related, economic and others) to predict, detect, suppress, prevent, and respond to information threats and mitigate their impact;

(e) information security forces are government bodies, as well as units and officials of government bodies, local authorities and organizations tasked to address information security issues in accordance with the legislation of the Russian Federation;

(f) information security means are legal, organizational, technical and other means used by information security forces;

(g) the information security system is a combination of information security forces engaged in coordinated and planned activities, and information security means they use;

(h) the information infrastructure of the Russian Federation (hereinafter referred to as the "information infrastructure") is a combination of informatization objects, information systems, Internet websites and communication networks located in the territory of the Russian Federation, as well as in the territories under the jurisdiction of the Russian Federation or used under international treaties signed by the Russian Federation.

Article 20 of the Doctrine identifies the strategic objective of ensuring information security in the field of national defence so as "to protect the

---

[318] Doctrine of Information Security of the Russian Federation, approved by the Decree of the President of the Russian Federation No. 646 of December 5, 2016, last retrieved on 19 December 2018 at http://www.mid.ru/en/foreign_policy/official_documents/-/asset_publisher/CptICkB6BZ29/content/id/2563163

[319] Ibid

vital interests of the individual, society and the State from both internal and external threats related to the use of information technologies for military and political purposes that run counter to international law, including for the purposes of taking hostile actions and acts of aggression that undermine the sovereignty and territorial integrity of States and pose a threat to international peace, security and strategic stability".[320]

Information security regulation is a rapidly developing area in Russia, catching up with technological breakthroughs. Among the latest normative documents there are Resolutions No. 747 of June 29, 2018, No. 772 of June 30, 2018, order No. 1322-R of 30 June 2018 on requirements to recording personal identification in biometric identification and authentication system; the approval of the draft Federal law "On amendments to article 51 of the Federal law "On Communication", which proposed to oblige state bodies and organizations, established to perform the tasks assigned to the Federal state bodies, to keep upto three years of data on the users of communication services operating the terminal equipment of these state bodies and the organizations, to keep these data up to date and to provide it in three-day time at the request of Federal Executive authority in the field of state protection.

Federal law N 187-FZ "On the Security of the Critical Information Infrastructure of the Russian Federation" dated 26 July 2017, sets out the framework and principles of safety of the critical information infrastructure of the Russian Federation, including the basis for the state's system of detection, prevention and elimination of consequences of computer attacks on the Russian information resources. It also set the mechanism of prevention of computer incidents at the Russian objects of critical information infrastructure.[321] This law will apply to companies operating information systems in the areas of healthcare, science, transport, communications, energy, banking and other financial market areas, the fuel and energy complex, atomic energy, defence, and the space, mining, metallurgy and chemical industries, as well as to persons that maintain the interaction of such systems. Among other things, the owners of critical information infrastructure facilities must: (i) immediately inform authorized bodies of computer incidents; (ii) assist authorized officials in detecting, preventing and eliminating the consequences of computer attacks; and (iii) ensure that an operating procedure is implemented for devices designed to detect, prevent and eliminate the effects of computer attacks.[322]

---

[320] Ibid

[321] Federal law N 187-FZ "On the security of critical information infrastructure of the Russian Federation" dated 26 July 2017, last retrieved on 21 October 2018 at http://government.ru/activities/selection/525/28727/

[322] Ibid

In 2016, Russia adopted the new version of the counter-terrorism laws, which imposed surveillance duties on almost any over-the-top internet communication services. It also established the first statutory requirement to provide information necessary to "decode" (essentially decrypt) encrypted electronic user communications to Russian security agencies upon their request. [323]

Russia requires all personal data operators that collect and process the personal data of Russian citizens to use databases located in Russia. Operators, when collecting personal data, must ensure that the databases used to record, systematise, accumulate, store, amend, update and retrieve them are located in Russia. Additional obligations are implemented for certain categories of online communication service, namely those that operate various chat and messenger services that enable their users to communicate with each other. The counter-terrorism amendments also provide for a number of other obligations.

Research

The national program "Digital economy of the Russian Federation" was launched in July 2017 in order to implement the Strategy for the Development of Information Society in the Russian Federation for 2017–2030 approved on 9 May 2017. The estimated amount of funding for 2019–2024 will exceed 1.8 trillion roubles (about 26 billion USD), of which more than 1 trillion roubles are the funds of the Federal budget.[324] The Federal program consists of six projects.

Table 2.5: National program "Digital economy of the Russian Federation" 2019–2024[325]

No	Project	Objective	Funding, approx.
1	On improving the regulation of the digital environment	Improving the regulation of the digital environment to develop a modern and flexible legal framework,	1.5 trillion roubles/22.1 billion USD

(Contd.)

---

[323] Federal Law dated 6 July 2016 No 374-FZ on changes to Federal Law on "Counter measures to Terrorism" last retrieved on 21 October 2018 at http://kremlin.ru/acts/bank/41108

[324] Заседание Правительственной комиссии по цифровому развитию, использованию информационных технологий для улучшения качества жизни и условий ведения предпринимательской деятельности/ Meeting of the Government Commission on Digital Development, Use of Information Technology to Improve the Quality of Life and Business Environment, 25 December 2018, last retrieved on 4 January 2019 at http://government.ru/news/35192/

[325] Meeting of the Government Commission on digital development, use of information technology to improve the quality of life and business environment, 28 December 2018, last retrieved on 4 January 2019 at http://government.ru/news/35192/

		unified requirements for electronic transactions, electronic documents, data storage, clear rules of application of advanced technologies in financial markets.	
2	Creation of information infrastructure for the digital economy	Create a global infrastructure for data transmission, storage and processing based on the Russian solutions. Continue to implement various digital platforms.	772 billion roubles/12 billion USD
3	Human resources for digital economy	Provide the digital economy with qualified personnel	143 billion roubles/2.1 billion USD
4	Information security	Major attention is paid to the Russian software, which can guarantee the protection of personal data, payment systems.	30 billion roubles/429 million USD
5	Creation of Russian end-to-end digital technologies	To bring them to international markets export volumes of our software and software solutions are relatively	450 billion roubles/6.5 billion USD
6	Digital public administration	Development of e-government services to expand access to public services online, in a "one-stop-shop" mode	235 billion roubles/3.3 billion USD

The Russian analogue of the Silicon valley[326], innovation centre "Skolkovo," launched in 2010, is a modern scientific and technological complex for the development and sale of new technologies, a science city near Moscow. With a strong political back up (Federal law of the Russian Federation No. 244-FZ on innovation centre "Skolkovo"[327]), its goal is to create a sustainable ecosystem of entrepreneurship and innovation. It includes five clusters: biomedical, nuclear, space, IT and energy and provides special tax conditions for companies working in these priority high technology sectors.

---

[326] About, Skolkovo Foundation, last retrieved on 4 January 2019 at http://sk.ru/foundation/about/

[327] Federal Law from 28.09.2010 No 244-FZ, 30 September 2010, last retrieved on 26 December 2018 at http://kremlin.ru/acts/bank/31792

It is planned that by 2020 about 50 thousand people will live and work on the area of 2.5 million square metres. With the public-private funding, the investment until 2020 will amount to 502 billion roubles (7.4 billion USD).[328]

> **Skolkovo IT R&D**[329]
> New systems of search, recognition and processing of audio, video and graphic information.
> Development of communication and navigation technologies
> New ways of storing, processing, transmitting and displaying information
> Development of new high-performance computing and data storage systems
> Apps and new IT technology
> Digital security
> Mobile, embedded, and wearable devices, as well as software and IoT applications
> Big data processing and analysis
> New developments in computer graphics and games
> Intelligent robotics and Autonomous vehicles
> Cloud technologies and services
> New systems for improving efficiency of production and business systems

Another correlation with the US reality, this time with DARPA—the Russian Foundation for Advanced Research in the Defence Innovations,[330] created in October 2012 (Federal Law No 174-FZ) and tasked to facilitate high risk defence and security research in order to obtain breakthrough results in military, technological and socio-economic areas. The foundation runs 71 laboratories and more than 50 working groups, which employ more than one and a half thousand highly qualified specialists. Among the current projects are the developments in information security, technologies for target detection and recognition, robotics, medical and biological research.

However, the funding situation of the Russian Foundation differs drastically from DARPA. In 2013 the budget was 3.3 billion roubles (48.8 million USD), and in 2014—3.8 billion roubles (56.2 million USD). The annual budget for 2015 and 2016 was initially planned to be about 4.5 billion roubles (66.5 millonn USD), but was reduced by 10 percent due to financial constraints.[331]

---

[328] Большую часть инвестиций в "Сколково" составят частные средства/Majority of investments to Skolkovo will be from private sources, 13 August 2013, last retrieved on 4 January 2019 at https://rg.ru/2013/08/13/skolkovo.html

[329] Skolkovo Foundation, Directions, last retrieved on 3 January 2019 at http://sk.ru/foundation/itc/p/directions.aspx

[330] Fund for Innovative Research, last retrieved on 3 January 2019 at https://fpi.gov.ru/

[331] Глава ФПИ: будет война операторов и роботов, а не солдат на поле боя/Head of FPI (Russian Foundation for Advanced Research in the Defence Innovations): the war will be between operators and robots, not between the soldiers, RIA News, 6 July 2016, last retrieved on 14 August 2018 at https://ria.ru/20160706/1459606647.html

---

Cyber events

In 2017, the number of crimes in the field of information and telecommunication technologies in Russia increased from 65,949 to 90,587, Russia Today was informed by the Russian Prosecutor General's office. The share of all criminal acts registered in Russia is 4.4%, which is almost every 20th crime. The most common cybercrime is illegal access to computer information (article 272 of the Criminal Code), as well as the creation, use and distribution of malicious computer programs (article 273 of the Criminal Code).

In July, Deputy Interior Minister Igor Zubov said that about 40 thousand crimes in the field of information technology have been committed in Russia since the beginning of 2018.[332]

At the meeting with representatives of the operational headquarters to ensure the security of the 2018 FIFA World Cup, President Vladimir Putin said that the security forces during the World Cup had a successful response and provided against about 25 million cyberattacks and other criminal attacks that threatened the information infrastructure of Russia.[333]

In April 2018, the National Association of International Information Security was set up in Russia. Its constituent founders are Lomonosov Moscow State University, the Moscow State Institute of International Relations (University) of the Russian Foreign Ministry, the Diplomatic Academy of the Foreign Ministry, the Academy of National Economy and Public Administration under the President of the Russian Federation, the Institute for Modern Security Challenges and the editorial board of Mezhdunarodnaya Zhizn ("International Life") magazine. The main goal of the association is to assist in the implementation of state policy in promoting international information security and advance Russian initiatives in this area. In addition, the association intends to participate in keeping civil society institutions in Russia and abroad informed and explaining to them the basic provisions of the Government's policy in this field.[334]

International Information Security Activity

For more than twenty years Russia has been the initiator of a number of specific activities in the UN and at other international organizations to

---

[332] Генпрокуратура рассказала о росте числа киберпреступлений в России/ The Prosecutor General's office told about the growth of cybercrime in Russia/, 14 August 2018, RT, last retrieved on 4 January 2019 at https://russian.rt.com/russia/news/545081-genprokuratura-rasskazala-o-roste-chisla-kiberprestupleniy

[333] Встреча с представителями штаба по обеспечению безопасности чемпионата мира по футболу 2018 года/ Meeting with representatives of the 2018 FIFA world Cup security headquarters, 16 July 2018, last retrieved on 4 January 2019 at http://kremlin.ru/events/president/news/58010

[334] Press release on the creation of the National Association of International Information Security, last retrieved on 4 January 2019 at http://www.mid.ru/mezdunarodnaa-informacionnaa-bezopasnost/-/asset_publisher/UsCUTiw2pO53/content/id/3169316

promote the creation of the international information security platform and strengthen all related areas.

In 1998, Russia was among the first to alert the United Nations on the upcoming challenge of the international information security threats. A special letter was sent by the Minister of Foreign Affairs of the Russian Federation to the UN Secretary-General in September 1998, accompanied by a draft resolution on "Developments in the field of information and telecommunications in the context of international security."[335] Emphasis was made on the need to halt an emerging—information—confrontation area and unlash new military conflicts. As a follow up, the Resolution 53/70 on "Developments in the Field of Information and Telecommunications in the Context of International Security"[336] was adopted at the UN General Assembly's Fifty-third session to solicit information security views and assessments from Member States.[337]

This resolution initiated the discussion of the new international regulation on information technologies, including on the issues of information security, definition and assessment of threats, prevention of military application of the new technologies. Since 1998, the Russian government has annually introduced a draft resolution in the First Committee on 'Developments in the field of information and telecommunication in the context of security'. With gradual changes, the non-binding resolution has been adopted by the UN General Assembly (UNGA) each year.[338]

The Shanghai Cooperation Organization (SCO), currently regarded as "Alliance of East", was announced in 2001 by China, Kazakhstan, Kyrgyzstan, Russia, Tajikistan, and Uzbekistan. It was established as a multilateral association to ensure security and maintain stability across the vast Eurasian region, join forces to counteract emerging challenges and threats, and enhance trade, as well as cultural and

---

[335] A/C.1/53/3, Letter dated 23 September 1998 from the Permanent Representative of the Russian Federation to the United Nations addressed to the Secretary-General, last retrieved on 4 January 2019 at https://digitallibrary.un.org/record/261158/files/A_C.1_53_3-EN.pdf

[336] UN Resolution A/RES/53/70 on Developments in the Field of Information and Telecommunications in the Context of International Security, United Nations, A/RES/53/70, 53rd sess. (1999), last retrieved on 14 August 2018 at http://www.un.org/ga/search/view_doc.asp?symbol=A/RES/53/70

[337] International cooperation in information security, 6 February 2004, last retrieved on 14 August 2018 at http://www.mid.ru/mezdunarodnaa-informacionnaa-bezopasnost/-/asset_publisher/UsCUTiw2pO53/content/id/486848

[338] Web-site of the NATO Cooperative Cyber Defence Center of Excellence, web-page on the "United Nations", last retrieved on 14 August 2018, https://ccdcoe.org/un.html

humanitarian cooperation.[339] Following the successful adoption in 2005 of the UN Resolution on the "Developments in the Field of Information and Telecommunications in the Context of International Security,"[340] the heads of SCO Member States released a statement on information security, endorsing the UN's approach.

The Russian information security concerns, among others, are the following[341]:

- ICT threats to internal affairs of sovereign states,
- military and political ICT uses that threaten international stability, and
- the on-going use of ICTs by some countries that adversely affect the whole world.

From the beginning a deep and constructive discussion of international information security in the UN First Committee has been challenged by principally different approaches to "information security" by the US and other liberal democracies on one side and the Shanghai Cooperation Organization (SCO) countries on the other. Disagreement extended to key definitions; exact scoping of the topic; threat perception as well as the mandate and role of the UN in general and the First Committee in particular in resolving international information security issues.[342]

In 2009, SCO Member States concluded the Agreement on Cooperation in the Field of Ensuring International Information Security (Yekaterinburg, 16 June 2009). On 9 January 2015, six members of the Shanghai Cooperation Organization (SCO) (China, Kazakhstan, Kyrgyzstan, Russia, Tajikistan, and Uzbekistan) proposed an updated version of the International Code of Conduct for Information Security to the United Nations, which was initially developed in 2011 by four members of the SCO - China, Russia, Tajikistan and Uzbekistan.

---

[339] R. Alimov, The Role of the Shanghai Cooperation Organization in Counteracting Threats to Peace and Security, UN Chronicle, Vol. 3, October 2017, last retrieved on 14 August 2018 at https://unchronicle.un.org/articlerole-shanghai-cooperation-organization-counteracting-threats-peace-and-security

[340] UN Resolution Developments in the Field of Information and Telecommunications in the Context of International Security A/RES/60/45 United Nations, 60th sess. (2006), last retrieved on 14 August 2018 at https://undocs.org/A/RES/60/45

[341] "Zayavleniye Glav Gosudarstv-Chlenov SHOS po Mezhdunarodnoy Informatsionnoy Bezopasnosti [Statement of Heads of SCO Member States on International Information Security]," The Shanghai Cooperation Organization, last modified 15 June 2006, last retrieved on 16 August 2018 at http://www.sectsco.org/RU123/show.asp?id=107

[342] Eneken Tikk-Ringas, Developments in the Field of Information and Telecommunication in the Context of International Security: Work of the UN First Committee 1998-2012, 2012, ICT4 Peace Publishing, Geneva, last retrieved on 17 August 2018 at https://ict4peace.org/wp-content/uploads/2012/08/Eneken-GGE-2012-Brief.pdf

---

International code of conduct for information security[343]
The purpose of the present code is to identify the rights and responsibilities of States in the information space, promote their constructive and responsible behaviours and enhance their cooperation in addressing the common threats and challenges in information space, so as to ensure that information and communications technologies, including networks, are to be solely used to benefit social and economic development and people's well-being, with the objective of maintaining international stability and security.

In 2018, 20 years after international security was placed on the UN's agenda in 1998, Russia keeps providing vast contributions to the international consensus in this area.

On 5 December 2018, the UN General Assembly adopted a Russian resolution on international information security (IIS) titled "Developments in the field of information and telecommunications in the context of international security."[344] It was supported by an overwhelming majority of states and co-authored by over 30 countries from different parts of the world. The breakthrough decisions aimed at strengthening IIS have been adopted in the anniversary year of 2018. These decisions aim to protect the interests of all countries in the sphere of cyber security regardless of their level of technological development.[345]

The document includes a number of innovations, starting with a provisional list of 13 international rules, norms and principles of responsible behaviour of states in this sphere. These actually form the world's first code of conduct in the digital sphere, designed to create the foundation for peaceful interaction and to prevent war, confrontation and any other aggressive action.[346] This document also includes the following provisions:

- Accusations of organizing and implementing wrongful acts brought against states should be substantiated.
- States must not use proxies to commit internationally wrongful acts using ICTs.

---

[343] Letter Dated 12 September 2011 From the Permanent Representatives of China, the Russian Federation, Tajikistan, and Uzbekistan to the United Nations Addressed to the Secretary-General, A/66/359, United Nations, 66th sess. (2011), last retrieved on 17 August 2018 at https://ccdcoe.org/sites/default/files/documents/UN-110912-CodeOfConduct_0.pdf

[344] A/C.1/73/L.27* Developments in the field of information and telecommunications in the context of international security, 22 October 2018, Seventy-third session, First Committee, last retrieved on 27 October 2018 at https://undocs.org/A/C.1/73/L.27

[345] Press release on the adoption of a Russian resolution on international information security at the UN General Assembly, 7 December 2018, last retrieved on 17 December 2018 at http://www.mid.ru/mezdunarodnaa-informacionnaa-bezopasnost/-/asset_publisher/UsCUTiw2pO53/content/id/3437775

[346] Ibid

- The United Nations should play a leading role in promoting dialogue among Member States to develop common understandings on the security of and the use of ICTs.
- State sovereignty and international norms and principles that flow from sovereignty apply to State conduct of ICT-related activities and to their jurisdiction over ICT infrastructure within their territory.
- States have a primary responsibility for maintaining a secure and peaceful ICT environment.[347]

Russia has also proposed convening an open-ended working group (OEWG) acting on a consensus basis for all UN Member States to take part in its activities. The OEWG will be authorized to consider the entire range of issues related to IIS. It will continue, as a priority, to further develop the rules, norms and principles of responsible behaviour of states in information space, study how international law applies to the use of ICTs by states and build up the digital capability of the developing countries. It is the first time that the UN group on IIS has been given such a mandate. Moreover, the status of discussions on IIS at the UN has been enhanced. Unlike the previous UN group of governmental experts on IIS, the OEWG will be a fully-fledged body of the UN General Assembly with the right to draft and recommend any documents to member states, including drafts of international treaties.[348]

2.9.3 Cyber Security in China

As per the Information Warfare Monitor issued in March 2009, China made cyberspace a key pillar of their national security strategies. China is actively developing an operational capacity in cyberspace, correctly identifying it as the domain in which it can achieve strategic parity, if not superiority, over the military establishments of the United States and its allies. Chinese cyber warfare doctrine is well developed, and significant resources have been invested by the People's Liberation Army and security services in developing defensive and offensive capabilities.[349]

In its efforts to turn the country from a "big cyber power" into a "cyber superpower," the Chinese leadership is active at three interconnected areas: information control, cybersecurity, and international norms and diplomacy.[350]

---

[347] Ibid
[348] Ibid
[349] Information warfare monitor, March 29, 2009, p. 7, last retrieved on 17 August 2018 at http://www.nartv.org/mirror/ghostnet.pdf
[350] A. Segal, Year in Review: Chinese Cyber Sovereignty in Action, 8 January 2018, last retrieved on 17 August 2018 at https://www.cfr.org/blog year-review-chinese-cyber-sovereignty-action

The guiding principle of the cyber regulations in China is the priority of the nation state's safety and security, which should be ensured through the monitoring and control for the use of cyberspace. China, through the Shanghai Cooperation Organization (SCO), introduced its strong position of cyber sovereignty together with Russia and several Central Asian countries, first in 2011 and later repeated in 2015.[351] Like Russia, China uses the term information security which covers the monitoring and control of the information flow, censorship of the content, but at the same time security of networks and computers.

According to the Chinese government, 802 million people are now actively using the internet, 57.7 percent of the population. This data was published by the China Internet Network Information Center (CNNIC) which is a branch of the country's Ministry of Industry and Information. By comparison, the United States has an estimated 300 million internet users.[352]

The National Informatization Plan issued in 2016 pledged to have an "impregnable" cyber system in 2025 by building up a group of globally competitive multinational companies and world-class mobile communications networks. According to the plan, the country should become a leader in some key technologies within information and communications by 2020. Breakthroughs in 5G-related technologies are expected at the same time.[353]

Highest Cyber Security Authorities

Founded in 2014, the Cyberspace Administration of China (CAC), executive body of the Central Cyberspace Affairs Commission, oversees implementation and enforcement of the Cybersecurity Law. The Commission is headed by the Communist Party General Secretary Xi Jinping. The deputy heads are the Premier of the State Council of China, and the Head of the Propaganda and Ideology Leading Group.

The National Computer Network Emergency Response Technical Team/Coordination Center of China (known as CNCERT or CNCERT/CC) was founded in September 2002. It is a cybersecurity technical centre and the key coordination team for China's cybersecurity incident detection, security evaluation, early warning and emergency response

---

[351] International Code of Conduct for Information Security. Report of the United Nations General Assembly, 13 January 2015, last retrieved on 17 August 2018 at https://ccdcoe. org/sites/default/files/documents/UN-150113-CodeOfConduct.pdf

[352] Nial Mccarthy, China Now Boasts More Than 800 Million Internet Users And 98% Of Them Are Mobile, 23 August 2018, last retrieved on 23 August 2018 at https://www. forbes.com/sites/niallmccarthy/2018/08/23/china-now-boasts-more-than-800-million-internet-users-and-98-of-them-are-mobile-infographic/#441b73587092

[353] Mandy Zuo, China aims to become internet superpower by 2050, 28 July 2016, last retrieved on 23 August 2018 at https://www.scmp.com/news/china/policies-politics/article/1995936/china-aims-become-internet-cyberpower-2020

community. CNCERT has branches and offices in 31 provinces, autonomous regions and municipalities across mainland China.[354]

Cyber Security Regulations

Standardization and control are the foundation of the information security in China. This has been reconfirmed by the new Cybersecurity Law of the People's Republic of China ("CSL")[355], effective as of 1 June 2017, adopted at the 12th People's Republic of China National People's Congress. The law became a platform for the state regulation of information technologies. It enforces the rights and obligations the government, network operators and users have in the area of cyber security and data protection; introduced increased control, strict responsibilities for the law violations, privileges to domestic goods and a ban on anonymity. It sets out general principles and measures to support and develop network security, including oversight, preventive measures and emergency response.

Article 1 of the Law states that this law is designed to ensure network security, protect the sovereignty of cyberspace and national security, defend social and public interests, protect the legitimate rights and interests of citizens, legal entities and other organizations in order to promote the healthy development of Informatization of the economy and society.

The CSL enactment has increased operating costs as "operators of critical information infrastructure" are legally obligated to store personal information and important business data in China, provide unspecified "technical support" to security agencies and pass national security reviews. Those critical areas include information services, transport and finance. Companies that store or provide internet data overseas without approval can be suspended or shut down and their business licence revoked. For example, Apple began storing its Chinese users' iCloud accounts at a new data centre in southern Guizhou province run by a local company and in order to comply with the new regulations Apple reported that it had transferred the cryptographic keys needed to unlock users' iCloud accounts to China.[356]

Critical network equipment and network security products must be tested for compliance with national standards and mandatory requirements, certified by institutions or tested for security before selling. These functions are assigned to the agencies involved in the field of the

[354] http://www.cert.org.cn/publish/english/index.html
[355] Cybersecurity Law of the People's Republic of China, 7 November 2016, last retrieved on 23 August 2018 at http://www.npc.gov.cn/npc/xinwen/2016-11/07/content_2001605.htm
[356] Stephen Nellis and Cate Cadell, Apple moves to store iCloud keys in China, raising human rights fears, 24 February 2018, last retrieved on 23 August 2018 at https://www.reuters.com/article/us-china-apple-icloud-insight/apple-moves-to-store-icloud-keys-in-china-raising-human-rights-fears-idUSKCN1G8060

Internet, and to the State Council of China, who compiles and publishes a list of key Internet equipment and products in the field of Internet security, as well as ensures the promotion of security certificates and testing, preventing the possibility of duplication.

The anonymity of users is excluded. While registering access to the Internet, to a social network, connecting a landline or mobile phone, providing the client with the services of publishing information or its transmission, when signing an agreement (on the provision of services), the client must provide a genuine identity card. At the same time the state is developing reliable means of identification and storage.

White Paper Outlines Potential Uses of AI[357]
In September, the China Academy of Information and Communications Technology (CAICT), a think tank under the Ministry of Industry and Information Technology (MIIT), published an artificial intelligence security white paper that outlined development areas describing how Beijing aims to use AI to automate censorship and control public opinion, and improve public security. CDT has translated excerpts of the report:
[...] Artificial intelligence can predict the development trajectory for Internet incidents, strengthen early warning capabilities as events evolve, pre-emptively intervene in and guide public sentiment to avoid mass online public opinion outbreaks, and improve social governance capabilities. Currently, the main system for monitoring public opinion online has built upon a pre-existing base of big data analytics with natural language processing, machine reading comprehension, and other related technologies to improve the system's intelligence level.

China believes that the big Internet companies operating in China will benefit from the enhanced cybersecurity, i.e. Alibaba, Baidu, Shanda Group, NetEase, Tencent, Sina, Tom, Sohu and 360. As well as the state, as in case of cyberattacks or control over the infrastructure of these companies, there is a danger of establishing control over the Chinese segment of the Internet and financial flows through Chinanet.

The Chinese government has issued close to 300 new national standards related to cybersecurity over the past several years. These standards cover products ranging from software to routers, switches, and firewalls. They contribute to making China an increasingly difficult market for foreign firms to operate, not just for selling to government or state-owned enterprise customers, but across the commercial market in China, spanning all sectors reliant on ICT infrastructure, from manufacturing to transportation.[358]

---

[357] White Paper Outlines Potential Uses of AI, 1 January 2019, last retrieved on 14 January 2019 at https://chinadigitaltimes.net/2018/11/white-paper-outlines-potential-uses-of-artificial-intelligence/

[358] How Chinese Cybersecurity Standards Impact Doing Business In China, 2 August 2018, Brief, Center for Strategic and International Studies, last retrieved on 23 August 2018 at https://www.csis.org/analysis how-chinese-cyber security-standards-impact-doing-business-china

The National Standards on Information Security Technology—
Personal Information Security Specification (GB/T 35273-2017) took effect
on 1 May 2018. The Standard requires transparency, specificity and fairness
of processing purpose, proportionality, security, risk assessment, and the
respect of individuals' rights to control the processing of information
about them.[359]

What is "critical information infrastructure" (CII)?
Article 31 of the Cyber Security Law provides a non-exhaustive list of selected critical
industries and areas whose information infrastructure would be regarded as CII,
including public communications, information services, energy, transport, water
conservancy, finance, public services, and e-governance etc., and more broadly, other
information infrastructure which may cause serious consequences if it suffers any
damage, loss of function, or leakage of data. The specific scope of "CII" is yet to be
formulated by State Council.

The new draft Regulations provide a scope in line with the Cyber Security Law
but further lists out additional industries and sectors whose network facilities and
information systems should be included in the scope of CII:

Industries: healthcare, education, social security and environmental protection;

Information networks: radio and television networks, and internet; service
providers providing cloud computing, big data and other large public information and
network services;

Scientific research and production: defence industry, large equipment industry,
petrochemical industry, and food and drug industry;

Media and news: radio stations, television stations, and news services.

The State Council will set up a "CII Identification Guideline" and industry supervisory
authorities are required to follow this CII Identification Guideline for the purposes of
identifying CII in their respective industries and sectors.[360]

As shown in the table below, the CSL's implementation requires
regulations, rules and guidelines and most of which are still in the pipeline
with their drafts being released in order to solicit public opinions.[361]

---

[359] China's Personal Information Security Specification: Get Ready for May 1, last retrieved
on 23 August 2018 at https://www.chinalawblog.com/2018/02/chinas-personal-
information-security-specification-getready-for-may-1.html

[360] Clarice Yue et al., China Cybersecurity Law update: Finally, draft regulations on
"Critical Information Infrastructure", July 2017, last retrieved on 23 August 2018 at

[361] Jet Deng and Ke Dai, China: New Challenges Ahead: How To Comply With Cross-Border
Data Transfer Regulation In China, 11 May 2018, last retrieved on 24 August 2018 at
http://www.mondaq.com/china/x/700140/Data+Protection+Privacy New+Challenges
+Ahead+How+To+Comply+With+CrossBorder+Data+Transfer+Regulation+In+China

Table 2.6: CSL's supporting regulations and guidelines

Category	Title	Legal status
Laws	CSL	Effective
Regulations and Rules	Measures for Security Assessment of Cross-border Transfer of Personal Information and Important Data ("Draft Security Assessment Measures")	In the pipeline
	Regulation on Security Protection of Critical Information Infrastructure ("Draft CII Regulation")	In the pipeline
Guidelines (National Standards)	Information Security Techniques—Personal Information Security Specification ("Personal Information Security Specification")	Effective
	Information Security Technology—Guidelines for Data Cross-Border Transfer Security Assessment ("Draft Security Assessment Guidelines")	In the pipeline

On 27 June 2018, China's Ministry of Public Security released for public comment a draft of the Regulations on Cybersecurity Multi-level Protection Scheme (MPS). It sets out the details of an updated Multi-level Protection Scheme, whereby network operators are required to comply with different levels of protections according to the level of risk involved with their networks. The Draft Regulation stipulates a wide array of investigative powers for MPS and sanctions for non-compliant companies, ranging from on-site inspection, investigation, and "summoning for consultation" to monetary fines and criminal liability.[362]

Enforcement actions are already taken by Chinese authorities. In July 2017, the CAC and other three departments have jointly initiated the special action of privacy policy review against ten notable domestic network product and service providers, including WeChat, Taobao, JD, AutoNavi, Baidu Maps, Didi Chuxing, Alipay, Sina Weibo, Umetrip and Ctrip. Afterwards, the review result was released to the public in September, and most of the investigated companies have taken corrective measures to improve their privacy policies, including explicitly stating the ways and purposes of collecting and using personal information and obtaining data subjects' prior consent.[363]

---

[362] 27 July 2018, last retrieved on 24 September 2018 at http://www.mps.gov.cn/n2254536/n4904355/c6159136/content.html Link in Chinese

[363] Jet Deng and Ke Dai, China: New Challenges Ahead: How To Comply With Cross-Border Data Transfer Regulation In China, 11 May 2018, last retrieved on 24 September 2018 at http://www.mondaq.com/china/x/700140/Data+Protection+Privacy/New+Challenges+Ahead+How+To+Comply+With+CrossBorder+Data+Transfer+Regulation +In+China

In the second quarter of 2018, the CAC and its local branches shut down 1,888 websites and 720,000 accounts for hosting illegal content.[364]

Starting 1 November 2018, police officers have the authority to physically inspect businesses and remotely access corporate networks to check for potential security loopholes. Police will also be authorized to copy information and inspect records that "may endanger national security, public safety and social order."[365]

In March 2017, the Ministry of Foreign Affairs and the Cyberspace Administration of China jointly issued the International Strategy of Cooperation on Cyberspace.[366] The International Strategy lists the principle of peace before sovereignty. The first steps in China's plan of future action include bilateral and multilateral discussions on confidence-building measures and work with others to prevent an arms race in cyberspace. Specifically, the document states that China supports the UN General Assembly to adopt resolutions regarding information and cyber security, and continues to facilitate and participate in the processes of the United Nations Governmental Groups of Experts (UNGGE) and other mechanisms to seek broader international understanding and support for this initiative.[367]

Cyber Security Budgets

China's national domestic security budget is believed to be publicly unavailable after 2013 (e.g. Reuters, 5 March 2014;[368] The China Quarterly, December 2017[369]).

---

[364] Nectar Gan, Cyberspace controls set to strengthen under China's new internet boss, 20 September 2018, last retrieved on 24 September 2018 at https://www.scmp.com/news/china/politics/article/2164923/cyberspace-controls-set-strengthen-under-chinas-new-internet?MCUID=4672f7018e&MCCampaignID=fb2d1c6c1f&MCAccountID=3775521f5f542047246d9c827&tc=5&stream=top

[365] Shan Li, China Expands Its Cybersecurity Rulebook, Heightening Foreign Corporate Concerns, last retrieved on 24 September 2018 at https://www.wsj.com/articles/china-expands-its-cybersecurity-rulebook-heightening-foreign-corporate-concerns-1538741732

[366] Adam Segal, Chinese Cyber Diplomacy in a New Era of Uncertainty, Aegis Paper No 1703, A Hooevr Institution Essay, 2017, last retrieved on 24 September 2018 at https://www.hoover.org/sites/default/files/research/docs/segal_chinese_cyber_diplomacy.pdf

[367] Ibid., chapter 4, Plan of Action, Section 2 (Rule-based Order in Cyberspace).

[368] Michael Martina, China withholds full domestic-security spending figure, 6 March 2014, last retrieved on 24 September 2018 at https://www.reuters.com/article/us-china-parliament-security/china-withholds-full-domestic-security-spending-figure-idUSBREA240B720140305

[369] Sheena Greitens, Rethinking China's Coercive Capacity: An Examination of PRC Domestic Security Spending, 1992–2012, 3 July 2017, last retrieved on 24 September 2018 at https://www.cambridge.org/core/journals/china-quarterly/article/rethinking-chinas-coercive-capacity-an-examination-of-prc-domestic-security-spending-19922012/FDC08F840E3479EDD5FE0BA1BEAA44A1

According to a survey by global consultancy firm PwC, Global State of Information Security Survey 2018 ("GSISS"), the average cybersecurity budget by surveyed companies in mainland China and Hong Kong is 23.5 percent higher than the global average, with a total average budget of 6.3 million USD per company.[370]

Plans of Action

The Great Firewall of China[371] (GFW) is the combination of legislative actions and technologies enforced by the People's Republic of China to regulate the Internet domestically. Its role is to block access to selected ("censored") foreign websites and to slow down cross-border Internet traffic. The effect includes: limiting access to foreign information sources, blocking foreign Internet tools (e.g., Google search, Facebook, Twitter etc.) and mobile apps, and requiring foreign companies to adapt to domestic regulations. Besides censorship, the GFW has also influenced the development of China's internal Internet economy by supporting domestic companies and reducing the effectiveness of products from foreign Internet companies. GFW operates under the "Golden Shield Project" by the Bureau of Public Information and Network Security Supervision.

In July 2017, China released its New Generation Artificial Intelligence Development Plan.[372] It is a document that indicates the vast scale of Beijing's priorities for and investments in AI. Its implementation is advancing throughout all levels of government. Although the future trajectory of its AI revolution remains to be seen, China is rapidly building momentum to harness the power of state support and the dynamism of private enterprises in a new model of innovation.[373]

AI has become a clear priority for Chinese leaders under the aegis of an agenda to transform China into a "nation of innovation." For the 13th Five-Year Plan (2016-2020), China's ambitions to transform itself into a superpower in science and technology are clear. In August 2016, the 13th Five-Year National Science and Technology Innovation Plan called for China to seize the "high ground" in international scientific development, launching a series of 15 "Science and Technology Innovation 2030 Megaprojects" that included both big data and intelligent manufacturing

---

[370] Chinese companies' spending on cybersecurity is almost a quarter more than the global average, 7 December 2017, last retrieved on 24 September 2018 at https://www.pwccn.com/en/press-room/press-releases/pr-071217.html

[371] The Great Firewall. The Art of concealment, 6 April, 2013, last retrieved on 24 September 2018 at https://www.economist.com/special-report/2013/04/06/the-art-of-concealment

[372] A Next Generation Artificial Intelligence Development Plan, 8 July 2017, last retrieved on 24 September 2018 at https://chinacopyrightandmedia.wordpress.com/2017/07/20/a-next-generation-artificial-intelligence-development-plan/

[373] Elsa Kania, China's AI Agenda Advances, 14 February 2018, last retrieved on 26 September 2018 at https://thediplomat.com/2018/02/chinas-ai-agenda-advances/

and robotics. The "Internet Plus" Artificial Intelligence Three-Year Action Implementation Plan was released in May 2016. In May 2017 the Ministry of Science and Technology announced the decision to add "AI 2.0" to that initial line-up as a 16th megaproject.[374]

Cyber Events

In March 2017, Tencent and other companies were told to close websites that hosted discussions on the military, history, and international affairs. In July, telecoms were told to crack down on "illegal" VPNs (in response, Apple was forced to remove VPNs from the China app store). A month later, the Cyberspace Administration of China announced new regulations further limiting online anonymity. The country also demonstrated new technological prowess, censoring photos in one-to-one WeChat discussions and disrupting WhatsApp. Parallel to the tightening on social media, Beijing has deployed facial and voice recognition, artificial intelligence, and other surveillance technologies throughout the country.[375]

The recent report released by the Chinese Cybersecurity Emergency Response Team (CN-CERT)—a division of the Cyberspace Administration of China—assessed the country's cyber threats landscape and forecasted emerging threats. In 2017, the Chinese National Vulnerabilities Database archived 16,000 security vulnerabilities, a 47.4 percent uptick from 2016. Smart devices gained much attention in this sphere. The total number of archived IoT vulnerabilities increased by 120 percent in 2017 and 27,000 smart devices fell prey to unauthorized remote control every day.[376]

Survey findings for mainland China and Hong Kong, showed customer records were the most commonly acknowledged target of security infractions, flagged by 46% of respondents. Financial loss (38%) and email compromise (36%) were the next most significant impacts cited by respondents.[377]

China continues hosting global cyber conferences; thus, the 2nd China Cybersecurity Summit 2018 brought together about 150 the most influential cybersecurity industry experts and executives to share the latest market information and discuss the emerging cyber threats, which refer to data security, cloud security, cyber intelligence and cloud computing, etc.

---

[374] Ibid
[375] A. Segal, When China Rules the Web, Foreign Affairs, 2018, September-October Issue, last retrieved on 26 September 2018 at https://www.foreignaffairs.com/articles/china/2018-08-13/when-china-rules-web
[376] Qiheng Chen, China's Cybersecurity Headache, 12 May 2018, last retrieved on 24 September 2018 at https://thediplomat.com/2018/05/chinas-cybersecurity-headache/
[377] Chinese companies' spending on cybersecurity is almost a quarter more than the global average, 7 December 2017, last retrieved on 24 September 2018 at https://www.pwccn.com/en/press-room/press-releases/pr-071217.html

From 2013, China has been organizing The China Cyber Security Conference and Exposition, the leading information security event, and the next one will be held in Beijing in June of 2019. It will attract more than 3,000 information professionals from government, private sector, academia and more than 60 media outlets. It has the theme of "Globalization of Cyber Security", and the Conference and Expo addresses the most up-to-date security and cyber security issues; from the latest trends, risks, strategies, technologies, including case studies and solutions. It is organized by SKD Labs, National Computer Virus Emergency Response Center, National Quality Supervision and Testing Center of Security Product for Network and Information system and Beijing Innovation Alliance.[378]

---

[378] Introduction, China Cyber Security Conference & Exposition 2019, last retrieved on 3 January 2019 at http://nsc.skdlabs.com/en/

3   Future Challenges

With the global expansion of cyber technologies and their constant evolution, the challenges associated with them are growing proportionally. Some of these can be predicted and prevented, while the others will be discovered retroactively. We are learning in this field as we progress, adjusting to and happily accommodating the previously unimaginable, appreciating the new helpful e-services and e-products. It is inhuman to halt this progress, but it is also inhuman not to halt the existential dangers it is bringing.

3.1 Global Interconnectivity and Internet Dependence

Forecast

- Enhanced technologies will keep stretching boundaries of global connectivity covering the entire planet.
- Computer devices, with cognitive platforms, will be reduced in size and the advanced means of connectivity, e.g., through wireless access points, satellites, etc., will decrease the burden of cable installations connecting continents.
- IoT will supply more international kinds of products and services, replacing ownership for rents and subscription.
- Further research in the applications of machine learning will resolve the global human need for using data accumulation, protection and analysis and ensure global accessibility.
- The Internet will be officially recognized as a critical infrastructure at the global level.
- Corporate and industrial systems will be isolated from the global network with a filtered connection with AI-empowered access controls.
- The threat of cyber conflicts will increase the risks of global propagation and be regulated through innovative technological and legal approaches for risk mitigation.
- Global connectivity will surpass the planet level and will be used for interplanetary signal transmission.
- National Internet subnetworks will become isolated on the infrastructural level.

Challenges

- Keeping the balance between global transparency and national sovereignty without jeopardizing progress.
- Monitoring and preventing the risks of global distribution of cyber conflicts, as well as mitigating the risks of cyber-crime and terrorism.
- Keeping the balance between restrictions and new technologies application.
- Monitoring cyber risks to critical infrastructure and CBRNe to be protected in the periods of technological transition through innovative solutions.
- The isolation of the Internet for civilian use and for the use of critical infrastructure.
- The threats will always evolve, and the existing laws will already be a response measure.
- Managing the dual-use nature of cyber tools and exercising cyber arms control (development, production, and acquisition).

3.2 Next Generation Cyber Attacks

Forecast

- Next generation AI-enhanced cyberattacks, accurately aimed, difficult to attribute, exploiting AI-based vulnerabilities, will have an evolving nature and self-improvement mechanisms.
- Spam, phishing, and spear-phishing techniques will involve machine learning for creation of fake emails or websites, targeting humans as a major vulnerability.
- More sophisticated and complex malware kits with advanced functionality will emerge.
- The rise of cryptocurrency for criminal use will create new services for black market cybercrime.
- Web applications will continue being the most vulnerable entrance point for attacks.

Challenges

- Developing a flexible, adaptive, self-learning defence system, including for AI vulnerabilities and AI versus AI attacks.
- Creating threat data throughout the global bank and research centre.
- Ensuring the vigilance of users and operators (human factor), and increasing personal and corporate hygiene.
- Sufficient investment in solutions and infrastructures which are secure by design and education for experts in relevant fields of study.
- Combatting APT attacks.

3.3 Next Generation of Cyber-physical Weapons

Forecast

- Autonomous weapon systems will become more advanced and independent in their decision making.
- Military activities will be implemented through cyber and cyber-physical operators.
- Human operators will take only final decisions (e.g., on lethal actions), the rest of the mission may be carried out by an AWS.
- Development and deployment of self-driving public transport and autonomous civilian infrastructure.
- New types of entertainment and sports, that involve AI-based art and cyber-physical systems

Challenges

- Preventing cyberattacks in remote communication, development, maintenance, and in the supply chain.
- Preventing erroneous fatal decision-making by AWS.
- Halting the cyber arms race.

3.4 Next Generation Warfare (Hybrid, 4GW)

Forecast

- The next level of hybrid war will spread across all the battlefields at once.
- Large-scale physical warfare operations will be reduced to the minimum deterrence level.
- 4GW, without any clear rules and procedures, concrete timelines, or centralisation, will blur the distinction between war and peace.
- Next generation of hybrid weapons will be enhanced by AI and will use AI sophisticated stealth malware.
- Next generation of autonomous robots and cyber-physical arms will replace human operators.
- Advanced cyber-physical attack techniques and operations, both in means and broader coverage.

Challenges

- Defence strategies will require a multi-domain approach and qualified warfare experts.
- Ensuring the safety of hybrid technologies from misuse, e.g., cyber enhanced nuclear technologies, cyber-enhanced biotechnical equipment, etc.

- Combatting the constantly evolving development of single-use exploits for targeted attacks.

> Weaponized human bodies based on advances in biohacking, subdermal implants and neural enhancement devices will be used in cyber terrorism and crime. Human performance enhancement will be also made through drugs and cognitive implants.

3.5 Cyber Regulations

Forecast

- The cyber arms environment will be conceptualized. New definitions developed for international regulations, including classification and clarification provided by cyber experts for offensive cyber tools.
- International cyber regulations developed and endorsed.
- Enforcement ensured, through the increased cooperation in digital investigation and evidence collection.
- More national regulations and frameworks will appear reflecting the national environment and international standards.
- New laws will appear based on ethical norms and standards and flexibly adjusting to new technology.
- Personal responsibility will be increased for any Internet activity, including malicious actions. Penalties and fines developed for minor cyber violations.
- Potential mandatory licensing and certification of cyber tools ensured.
- Global Cyber Security Index will keep measuring a country's commitment to cyber security.

Challenges

- Ensuring the flexibility of law and adaptability to emerging threats, and introducing new legal mechanisms, including ethical background regulations.
- Translating the outcomes of discussions on the development of internationally acceptable cyber regulations into reality.
- Drawing a conceited conclusion in discussion on the applicability of the international law to cyber related offensive operations.
- Developing a clear legal definition of "defence", and its scope and limits for cyber space.
- Harmonizing the nations approaches to developing cyber regulations and developing national frameworks.
- Regulating AWS autonomous decision making and missions developing a classification of offensive tools and corresponding legal responsibilities.
- Creating EWARs and implementing cyber related risk assessments.

1. "If we encounter a sentient AI, do we consider it a fellow citizen? Does it have human rights? Should it follow human laws and morals?"
2. If one wants to delete a sentient AI, can it be classified as a murder?
3. If the AI hacks something somehow, is it a crime? And if so, how do you prosecute it?

4   Conclusion

"Where there is great power there is great responsibility."

Winston Churchill

In less than a century our civilization has made an unprecedented cyber leap in its technological development. It is only fair to expect that the benefits will ensure a more accomplished lifestyle, guaranteed safety and security, and ultimate prosperity of humanity. However, we are facing an overwhelming multitude of new and unknown challenges and threats, and there is still a long way to go to a technologically enhanced Eden.

Invention of the boundless boundaries of cyberspace has been followed by its criminalization and militarization, with software codes used as cyber arms and agents in espionage and sabotage. Physical weapons, enhanced by computerized command and control, have acquired full autonomy and the ability to take decisions and select targets. The dual use character of cyber arms made them accessible to anyone and cyber markets are thriving. The constantly escalating cyber arms race is now our reality and the cyber powers invest billions in the competition for AI-enhanced superiority.

Hybrid war, a new type of war, exploiting all five military domains, has emerged, and its major characteristics are covert and speedy cyber operations, absence of clear protocols, unpredictability and uncertainty, as well as difficulties in attribution. Air, naval and ground autonomous weapon systems, combat robots, remotely-piloted systems (aircrafts, drones), self-targeting cruise missiles, submarine-hunting underwater unmanned vessels, and UGV units constitute the deterring arsenal of the cyber-age warfare. It employs all technological achievements and blurs the lines between peace, crisis, and conflict.

Nuclear weapons, enhanced by intelligent systems with sophisticated logic, have increased the danger exponentially. Development and modernisation of nuclear cyber systems have provided the greatest deterrent of modern times, as well as the greatest danger of existential level. Recognising the need of the nuclear systems, existence and continuing production of nuclear weapons and their support equipment, the issue of their protection from sophisticated cyber threats is becoming of paramount importance.

Constantly evolving attack vectors, entailing disrupted services, financial and reputational losses, are keeping cyber and forensic experts alert, and the multi-disciplinary research community is developing proactive defence measures and sophisticated security technology to protect computer systems and networks, especially with critical infrastructure and CBRNe. However, the transnational reach of cyberspace, and the covert and stealth nature of attacks, attribution challenges and engagement of high level and well-funded actors, prevent the task of ensuring a safe and secure cyberspace environment.

The security of the information space, and maintaining its peace and stability, was raised high on the international agenda and multiple initiatives have been proposed to come up with internationally applicable cyber regulations.

The Budapest Convention (2001), the Tallinn Manuals (2013 and 2017), Code of Conduct for Information Security proposed by the Shanghai Cooperation Organization (2015), reports of the UN Group of Governmental Experts on Information, Telecommunications, and International Security (2004–2017), UN Resolutions (1998–2018) reflect extensive activities in outlining the global cybersecurity landscape, identifying existing and emerging threats, capacity and confidence building, and developing recommendations on the implementation of norms, rules and principles for the responsible behaviour of States, and the application of international law to the use of information and communications technologies, etc.

Since 2007, the "ITU Global Cybersecurity Agenda", a framework for international cooperation, has been enhancing confidence and security in the information society. The Global Cybersecurity Index, the UN multi-stakeholder initiative to measure the commitment of countries to cybersecurity, evaluates their status within legal, technical and organizational measures, capacity building and cooperation, and implements the monitoring and reporting of cyber capabilities of countries.

In spite of all the undertaken efforts, there are a lot of issues to be agreed upon. They include applicability of international law to cyber conflicts resolution, cyber environment definitions, including the notorious dual use nature of cyber arms, scope and limits of self-defence and countermeasures, attribution and investigation ethics, international normative guidance and policies, etc. Meanwhile the countries develop bilateral and multilateral agreements, based on their best practices and balancing between their national sovereignty and global openness while aiming not to jeopardize progress.

In view of the above challenges and building on the extensive review of the real-life cyber events, it seems to us that it is legitimate to propose embedding an ethical background in the development of the next generation's laws for issues related to high technologies. The ethical norms of society should follow the skyrocketing technological advancements, if not to outpace them. It will take time and efforts to reach the rewarding Eden, much desired by our genetic memory, but this time enhanced by flourishing digital reality.

We hope this book will contribute not only to the peaceful use of cyberspace, but also to the peace and prosperity of the planet.

Glossary

The magnitude of the threat and the lack of globally accepted international regulations in the production and distribution of commercial and open source malicious software and techniques, i.e. cyber arms, urge for concerted action of the global cyber security community and legal experts. Conceptualization of the cyber environment, clear and explicit definitions and classifications provided by cyber professionals are essential for international legal experts engaged in the development of cyber laws and legal standards.

Over the last few decades, the field of cyber security has already established a set of terms, which are widely used. For the matter of consistency, the following definitions, in their alphabetic order, are provided for selected and ambiguous terms which are used in this book.

Advanced Persistent Threat (APT): APT is a set of sophisticated and multi-directional long term cyberattacks. It requires a high degree of covertness, expert knowledge and sufficient funding over a long period of time.

Artificial Intelligence (AI): AI is software able to rely on and take decisions based on the provided knowledge rather than on the predefined algorithms.

Autonomous Weapon Systems (AWS): AWS is a weapon system that, once activated, can select and engage targets without further intervention by a human operator.[1] AWS may operate in the air, on land, on water, under water, or in space.

CBRNe Cyber Security: A branch of cyber security, that ensures the safety of the chemical, biological, radiological, nuclear and explosive (CBRNe) equipment and facilities.

Cyber Arms: A cyber arm is software able to function as an independent agent and run commands. It has a dual use nature and can be used both for offence and defence purposes.

---

[1] Department of Defense Directive, Number 3000.09, November 21, 2012. Incorporating Changed 1, May 8, 2017, Autonomy in Weapon Systems, last retrieved on 15 April 2018, http://www.esd.whs.mil/Portals/54/Documents/DD/issuances/dodd/300009p.pdf

Cyber Conflict: A cyber conflict is any conflict between two or more state or non-state actors in cyberspace.

Cyber Arms Industry: The cyber arms industry is the design, development, production, distribution, and acquisition of cyber related products and services, e.g., cyber arms, vulnerabilities, exploits, surveillance technologies, etc.

Cyber Attack: Is a one-sided offensive activity on the computer system or network to violate confidentiality, integrity, and availability of information, disrupt services, illegally use or destroy the computer system or network.

Critical Infrastructure: Critical Infrastructure defines assets that are essential for the functioning of a society including national safety and security. It includes power, water and heat generation and supply, ports, airports and ground crossings, hospitals, strategic information centres and research institutions, early warning systems and controls, etc.

Cyber-physical Systems: Cyber-physical systems are physical platforms, semi or fully autonomous, that are controlled by computerised control systems.

Cyber Resilience: Cyber resilience is the entity's ability to continue uninterrupted operations under cyberattacks.

Cyber Security: Cyber security is the protection of cyberspace, i.e. software, hardware, data stored in computer systems, as well as data in procession or transfer from one computer system to another.

Cyberspace: Cyberspace is a "global domain within the information environment consisting of interdependent networks of information technology infrastructures and resident data, including the Internet, telecommunication networks, computer systems, and embedded processors and controllers".[2]

Cyber Tools: Cyberarms used for specific purposes in cyber activities. They include information gathering tools, scanning tools, exploitation tools, assault tools, etc.

Cyber War (Cyber Warfare): Cyber war is an escalated cyber conflict between two or more parties with the purpose towards establishing dominance over an opponent or opponents in critical areas (political, industrial, financial, information, etc.). It can be a cyberwar *per se* or a part of offensive activities in other domains, i.e. a hybrid war.

---

[2] Department of Defense Dictionary of Military and Associated Terms, 8 November 2010 (as amended through 15 February 2016), last retrieved on 5 November 2018 at https://fas. org/irp/doddir/dod/jp1_02.pdf

Cyber Weapon: A cyber weapon is a cyberarm employed for military purposes; it is an offensive tool for cyber offensive operations.

Dual-use Technology: Technology is said to be of dual-use when it can be used for peaceful and malicious purposes.

Exploit: A constructed command or software designed to take advantage of a flaw in a computer system (vulnerability), typically for malicious purposes, such as accessing system information, establishing remote command line interface, causing denial of service, etc.

Malware: Malicious software specifically designed to gain unauthorized access to a computer system, disrupt operations or cause damage.

Vulnerability: This is a flaw in the software, hardware or networks causing it to be exposed to a physical or cyberattack.

Weapons of Mass Destruction (WMD): Following the first WMD definition provided by the UN Commission on Convention Armaments (CCA) in 1948, WMD are defined as "atomic explosive weapons, radioactive material weapons, lethal chemical and biological weapons and any weapons developed in the future which might have characteristics comparable in destructive effect to those of the atomic bomb or other weapons mentioned above".[3]

---

[3]   CCA, UN document S/C.3/32/Rev.1, August 1948, as quoted in UN, Office of Public Information, The United Nations and Disarmament, 1945–1965, UN Publication 67.I.8, 28. Also, "Resolution Defining Armaments," State Department Bulletin, August 29, 1948, 268.

Index